Lee's Building Maintenance Management

Fourth edition

Paul Wordsworth
MA, ARICS, MIMBM

b

Blackwell Science

Blackwell Science Ltd
Editorial Offices:
Osney Mead, Oxford OX2 0EL
25 John Street, London WC1N 2BS
23 Ainslie Place, Edinburgh EH3 6AJ
350 Main Street, Malden
    MA 02148 5018, USA
54 University Street, Carlton
    Victoria 3053, Australia
10, rue Casimir Delavigne
    75006 Paris, France

Other Editorial Offices:

Blackwell Wissenschafts-Verlag GmbH
Kurfürstendamm 57
10707 Berlin, Germany

Blackwell Science KK
MG Kodenmacho Building
7–10 Kodenmacho Nihombashi
Chuo-ku, Tokyo 104, Japan

First published in Great Britain by
Crosby Lockwood Staples 1976
Second edition published by
Granada Publishing Ltd 1981
Reprinted with minor updating revisions 1983
Third edition published by
William Collins Sons & Co. Ltd 1987
Reprinted by Blackwell Science 1988, 1992,
1993, 1995, 1998
Fourth edition published by
Blackwell Science 2001

Set in 10.5/13 Times
by Sparks Computer Solutions Ltd, Oxford
http://www.sparks.co.uk
Printed and bound in Great Britain by
MPG Books Ltd, Bodmin, Cornwall

DISTRIBUTORS
Marston Book Services Ltd
PO Box 269
Abingdon
Oxon OX14 4YN
(*Orders:* Tel: 01235 465500
          Fax: 01235 465555)

USA
Blackwell Science, Inc.
Commerce Place
350 Main Street
Malden, MA 02148 5018
(*Orders:* Tel: 800 759 6102
                781 388 8250
           Fax: 781 388 8255)

Canada
Login Brothers Book Company
324 Saulteaux Crescent
Winnipeg, Manitoba R3J 3T2
(*Orders:* Tel: 204 837-2987
           Fax: 204 837-3116)

Australia
Blackwell Science Pty Ltd
54 University Street
Carlton, Victoria 3053
(*Orders:* Tel: 03 9347 0300
           Fax: 03 9347 5001)

A catalogue record for this title
is available from the British Library

ISBN 0-632-05362-3

Library of Congress
Cataloging-in-Publication Data
Wordsworth, Paul.
    Lee's building maintenance management/
    Paul Wordsworth – 4th ed.
        p. cm.
    Includes bibliographical references and
index.
    ISBN 0-632-05362-3
    1. Buildings – Maintenance – Management. I.
    Wordsworth, Paul. II. Title.
TH3361 .L43 2000
658.2'02–dc21

                                00-058598

For further information on
Blackwell Science, visit our website:
www.blackwell-science.com

Contents

Preface

In the 25 years since this book was first published there have been many changes in the construction and property management sectors, not least of which has been the growing recognition of the role of the building maintenance manager. Building maintenance now accounts for over half the building industry's total output, and for over two thirds of the contracts let. Changes in social attitudes over this period have resulted in a much greater emphasis on environmental management, health and safety, and user-oriented service provision; and in this context the role of the maintenance manager continues to expand, as more demands are made by users regarding the economic and functional efficiency of the buildings in which they live and work.

Maintenance management as a professional discipline has developed apace during this period. Professionally accredited degrees are being offered at a number of universities and these, with other related technical and professional qualifications that focus more specifically on maintenance management practice as part of their syllabuses, have provided new pathways into building maintenance management.

In this regard Reginald Lee's book has played a significant role in underpinning the development of a philosophical and educational framework for building maintenance management, by providing a systematic and inclusive approach to the variety and range of issues, from the legal to the technical to the fiscal, which need to be co-ordinated by the modern maintenance manager.

In the 14 years since the last (third) edition, there have been many developments and a few paradigm shifts (for example in information technology) which require a new approach; but in many other areas, such as fiscal planning, I have retained and revised that considerable amount of the original text which continues to be true. It is my intent, as was Reginald Lee's, to provide a textbook which both underpins the academic courses and modules in building maintenance management, and also provides the experienced maintenance manager with a continuing professional development reference and an updated overview of new developments in the area.

The overall structure and sequence of the book has broadly been retained; though there has been some reorganisation of content within this framework. There are some changes in this new edition to the level of specific detail given in the text, for example in the areas of law and information technology. I have provided an overview of the most important areas and developments, together with a navigational guide as to where to find currently accurate detail in these fields. There is also a new chapter on the increasingly important areas of conservation and the environment, and I have given a more scientific approach to the section on defect diagnosis. There are, however, many other areas which could be expanded *ad infinitum* if space permitted. Consequently this book should be regarded not as an encyclopaedia of everything the maintenance manager may ever need to know, but as a structured and systematic framework for understanding and co-ordinating the mass of knowledge and information the modern maintenance manager is required to handle.

Building maintenance management, long an unrecognised area of professionalism, is at last coming to be viewed more generally for what it is: an essential activity to ensure the ongoing provision of the safe, comfortable, and affordable environment in which we all expect to live, work, and play. I hope this book continues to support the building maintenance management profession.

In the preparation of this book I must acknowledge help from a number of sources in terms of information, illustration, and advice. In particular I am indebted to Paul Hodgkinson for his graphic skills in preparing the illustrations and diagrams; to Building Maintenance Information Ltd for the use of their statistics, to Simon Brown for his help on the section dealing with multiskilling of maintenance operatives, and to many colleagues at Liverpool John Moores University for their help and advice.

Paul Wordsworth
Wirral

Chapter 1
Maintenance in Context

1.1 The significance of building maintenance

Building maintenance is big business. It accounts for over 5% of the UK's gross domestic product (GDP); in fiscal terms, over £30 billion per year[1], making it one of the largest industries in the UK economy. It is an essential activity, which supports lifestyle and livelihood; and which maintains the considerable asset value of the country's property stock, worth about £2000 billion at current prices[1], equivalent to over half the nation's total wealth. Yet because it is a diffuse operation, taking place incrementally through time, in many locations, and by many different organisations, the scale and importance of building maintenance work is frequently undervalued in comparison with higher-profile and more visible new construction. Building maintenance management is also a highly complex sphere of operations, involving the interaction between the technical, social, legal and fiscal determinants which govern the use of buildings. Indeed, it is increasingly true that building maintenance management is as much about providing a level of service to building users as it is about the buildings themselves, and in this respect the modern maintenance manager will have to rely as much on knowledge of the managerial and social sciences as on the traditional technical knowledge base of building construction and deterioration. Given the size and complexity of building maintenance management as an industry, it is perhaps surprising that its public profile as a profession and career remains comparatively low, though this perception may change as a result of the new orthodoxies of health and safety, conservation, and environmental protection, which may serve to focus and heighten the positive role good maintenance management may play in these areas.

The condition and quality of buildings is one of the most fundamental components of the quality of life. The vast majority of people spend over 95% of their time in or next to a building of one kind or another, so in this sense the built environment has become our 'natural' environment. The condition and quality of buildings reflect public pride or indifference, the level of prosperity in the area, social values and behaviour and all the many influences both past and present which combine to give a community its unique character. There can be little doubt that dilapidated and unhealthy buildings in a decaying environment depress the quality of life and contribute in some measure to antisocial behaviour.

Unfortunately these social consequences are difficult to quantify on a balance sheet and as a result are rarely given proper consideration.

In most cases maintenance decisions are based on expediency and over a period of time represent a series of *ad hoc* and unrelated compromises between the immediate physical needs of the building and the availability of finance. There is a lack of precise knowledge of the benefits which accrue from different levels of maintenance expenditure and little attempt is made to forecast the overall long-term effects of doing or failing to do work in this field. The reason may be that from the standpoint of the individual firm the amount spent on maintenance appears small in comparison with other operating costs. But when viewed on a national scale it is quite clear that maintenance is an activity of primary importance. Figure 1.1 gives the actual spend on building maintenance in the UK over a period of 16 years.

However, available evidence would suggest that in general buildings are under-maintained and that a substantial part of the building stock is in danger of deteriorating below the point of economic repair. The *English House Condition Survey 1981*[2] revealed that of the 18.1 m. dwellings in England around 2 m. were in poor condition and of these over 1 m. required repairs costing in excess of £7000 (at 1981 prices). Further evidence of the dilapidated state of the housing stock was contained in a report published in 1986 by the Audit Commission on Managing the Crisis in Council Housing which estimated that 85% of the 5 m. council houses were in need of repair and improvement. This shows little change from the situation reported in 1970 by the National Institute of Economic and Social Research[3] which concluded that the arrears of housing maintenance probably amounted to eight or nine times the volume of work actually carried out. However, the *English House Condition Survey 1996*[4] indicated that there has been some improvement in recent years, with 7.5% of the

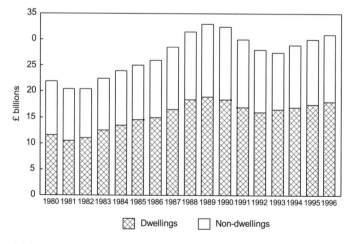

Fig. 1.1 Maintenance expenditure (£bn 1990 constant prices). (Source: *Building Maintenance Information Special Report*, December 1997.)

stock currently regarded as unfit. Improvements were noted particularly in the increasingly large registered social landlord (Housing Association) sector; although on average the repair backlog was £1500 per dwelling across the country. From the maintenance manager's point of view, the notable statistic was that privately owned houses were in a considerably worse state of repair than equivalent public sector (managed) houses. Although there are no comparable figures for nonresidential buildings it is probable that they suffer from a similar degree of care and neglect as the housing sector. Figure 1.2 gives average maintenance costs across a range of property and estate types.

1.2 Satisfaction of building needs

In order to put the problem into perspective it is necessary to view maintenance in the context of the overall building process. The building needs of the community are met by the interrelated construction activities of maintaining, modernising and replacing the existing stock of buildings and by the erection of additional new buildings (Fig. 1.3).

A frequently expressed view is that the level of expenditure on maintenance is too high. However, in order to judge the reasonableness of the expenditure it is necessary to consider three factors – firstly, whether the amount spent is excessive in relation to the work done, secondly, whether the work which is done is necessary and unavoidable, and thirdly, whether it would be advantageous to

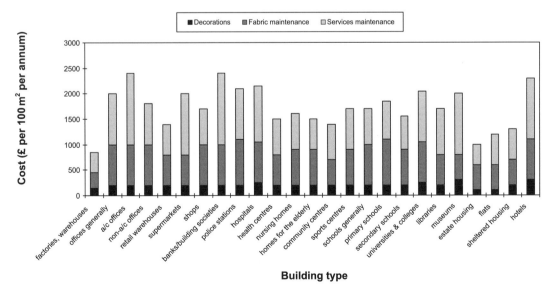

Fig. 1.2   Average maintenance costs 1998. (Source: *Building Maintenance Information Report 272*, November 1998.)

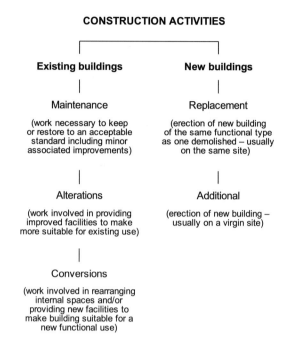

Fig. 1.3   Construction activities within the overall building process.

carry out more work. It would seem that the views expressed are well founded in relation to the first point and that better management and work planning would result in a more economic use of resources and a corresponding reduction in total costs. In regard to the second point there are many reported cases of early maintenance which could have been avoided by better design and the choice of more suitable materials.

The third point concerning the desirable level of maintenance cannot be considered in isolation. Clearly construction resources are limited and the object should be to achieve the optimum allocation of manpower, materials and capital between the maintenance and improvement of existing buildings and the construction of new buildings. Maintenance by arresting decay extends the physical life of a building and thereby delays replacement and defers expenditure on new construction. Of course, it can only be regarded as a substitute for new construction if the building remains functionally satisfactory.

To a large extent this will depend upon the degree to which it is possible to modify the internal layout to accommodate changing user requirements. There is thus a relationship between the adaptability of the design, building life, maintenance costs and new construction costs. Generally the more adaptable the building, the longer the period of use, resulting in higher total maintenance costs but lower new construction costs. In terms of assessing this balance, a cost-benefit equivalence may be drawn up as indicated in Fig. 1.4.

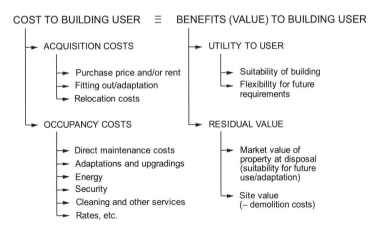

COST TO BUILDING USER   ≡   BENEFITS (VALUE) TO BUILDING USER

→ ACQUISITION COSTS

    → Purchase price and/or rent
    → Fitting out/adaptation
    → Relocation costs

→ OCCUPANCY COSTS

    → Direct maintenance costs
    → Adaptations and upgradings
    → Energy
    → Security
    → Cleaning and other services
    → Rates, etc.

→ UTILITY TO USER

    → Suitability of building
    → Flexibility for future
      requirements

→ RESIDUAL VALUE

    → Market value of
      property at disposal
      (suitability for future
      use/adaptation)

    → Site value
      (− demolition costs)

Fig. 1.4   Costs and value of a building.

It will be noted from this equation that the utility to the user is a key variable which will affect whether a building remains viable, and hence the proportion of resources allocated to either maintenance of the existing building or procuring a new building. Levels of maintenance are set therefore according not only to the physical state of disrepair of the building, but are closely related to its actual use. Brand[5] makes the point that buildings in a state of repair no longer suitable for one user may find another user which does not need the standard of facilities of the first. This operation of the market is limited at the lower end by statutory health and safety, and fitness standards (see Chapter 3) which define the minimum acceptable standard for human occupation, depending on use.

For the most part, building labour can be used indifferently for either maintenance or new construction, and thus policies which influence the volume of new construction will also affect the availability of labour for maintenance. Stone[6] has argued that more output would be obtained from the same labour force by giving priority to maintenance and improvement, in that this would not only increase the rate at which satisfactory dwellings are made available but would also reduce the cost. Needleman[7] also emphasises the advantages to be gained from rehabilitation rather than wholesale redevelopment; a policy which was implemented by the Housing Act 1969, by providing much more generous grants for the improvement of houses which lack the basic amenities or which fall below an acceptable standard. However, if policies are adopted which by persuasion, financial inducement or otherwise result in an increase in the demand for maintenance it becomes even more important that this work should be properly planned.

1.3 The building process

The role played by maintenance in the overall building process is shown in

Fig. 1.5. The extent to which maintenance is considered at the design stage is likely to depend upon whether or not the owner, or the person commissioning the building, will be the subsequent user. Where the owner is a developer who intends selling or leasing the building on completion it is probable that maintenance will be considered only in so far as it is likely to affect the sale price or rent.

Even where the owner and user are one and the same, it is probable that the initial capital costs of the building and the subsequent maintenance costs will be drawn from different sources and not related. Also it is misleading to think of the owner and user as being individuals, in that for a building of any size they would constitute a complex system of differing interests; often a good deal of abortive work results from failure to recognise and resolve the conflict of interests and objectives. However, both new construction and maintenance are usually subject to financial constraints and, notwithstanding the difficulties, an attempt should be made to achieve a proper balance between the costs of these activities by analysing the costs-in-use of alternative design solutions[8]. Where the design requirements are stated in the form of performance specifications these could be used for establishing maintenance standards and incorporated in a maintenance manual together with other relevant detail concerning the construction of the building and its services. Figure 1.6 illustrates the life cycle of a building, and the various options and decisions which are made during the course of its life.

Eventually a decision will have to be made as to whether to demolish the building and replace it with a new one or to improve or adapt it to make it more suitable for either the present use or a new use. In such cases it is helpful to distinguish between physical life, functional life and economic life. Provided the structure is basically sound it is possible with proper maintenance to extend

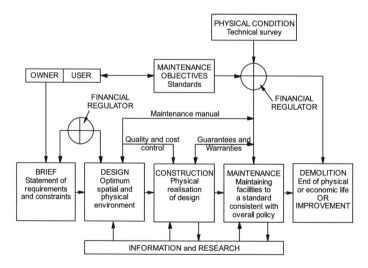

Fig. 1.5   The total building process, showing the role played by maintenance.

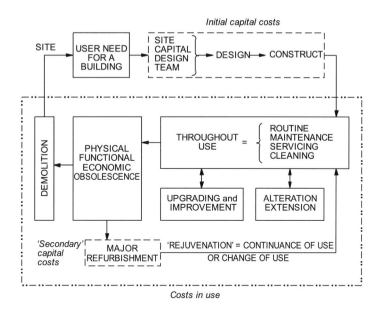

Fig. 1.6   Building life cycle.

the physical life of a building almost indefinitely. In the course of time it will tend to become increasingly unsuitable for the functional purpose for which it was originally designed. It may then be possible to adapt it for another use, which will usually involve a change of ownership and perhaps extensive alterations. Thus a building may have a series of different functional lives before it reaches the end of its economic life.

In some cases the economic life may be limited by a substantial rise in the value of the site on which the building stands. This may dictate the replacement of the building with one which will exploit the potential of the site to greater advantage. The determining factor will be the extent to which the value of the replacement building exceeds the cost of building it and the value of the present building. Switzer[9] describes economic life as 'that period of effective life before replacement; replacement will take place when it will increase income or reward absolutely, but will not reduce the rate of interest at present enjoyed from the landlord's total investments'. He advocates the control of the economic life of buildings by means of adaptability in the structure and by the judicious use of planning controls to regulate site values and thereby create the correct conditions for preservation or redevelopment. This would permit a closer relationship to be achieved between design durability and effective life.

Other factors that would have to be considered include social acceptance and the effects of new technology and legislation.

1.4 Maintenance defined

The definition of what constitutes building maintenance may seem at first sight self-evident, though on investigation it can be seen that there are several areas requiring closer attention; for example, the boundaries between maintenance and improvement, or the question of whether maintenance applies to an object or to a function.

BS 3811:1964 defines maintenance as 'a combination of any actions carried out to retain an item in, or restore it to, an acceptable condition'. The actions referred to are those associated with initiation, organisation and implementation. There are two processes envisaged: 'retaining', i.e., work carried out in anticipation of failure, and 'restoring', i.e., work carried out after failure. The former is usually referred to as 'preventive maintenance' and the latter as 'corrective maintenance'. There is also the concept of an 'acceptable standard'. This may be construed as acceptability to the person paying for the work, to the person receiving the benefit or to some outside body with responsibility for enforcing minimum standards. It can also be construed more widely as acceptability to the public at large or to specific sections of the public. Clearly there are no absolute standards which would be equally acceptable to everybody or which would remain acceptable to the same group of people over a period of time.

The standards acceptable at the time of undertaking the work may be higher or lower than the initial design standards. In many cases the standard deemed acceptable would be higher than that originally provided and the work would include an element of improvement. It could be argued that this interpretation would be inconsistent with the phrase 'to retain or restore' in that this would suggest the initial standard as the proper basis. However, with the passing of time buildings are modified to accommodate new uses and it becomes increasingly unrealistic to think in terms of keeping or restoring the initial standards.

In recognising the desirability of including a reasonable element of improvement, the Chartered Institute of Building[10] offers the following definition: 'Work undertaken in order to keep, restore or improve every facility, i.e., every part of a building, its services and surrounds, to an agreed standard, determined by the balance between need and available resources'. This introduces the notion of an 'agreed standard' which, from the general tenor of the definition, is assumed to be higher than the initial standard. There may, of course, be cases where buildings are put to a less demanding use for which lower standards would be acceptable. There is also reference to 'the balance between need and available resources', which would be an important factor to take into account when fixing an acceptable standard. The effects on both the value in use and the value on sale or letting would have to be considered. There are difficulties in relating the value on sale to the standard of maintenance, in that the market value of property is determined by many factors other than physical condition. Commercial properties in particular show a tendency to appreciate in value as a result of surrounding development rather than the attributes of the building.

The reference to available resources is interesting in that it suggests that some arbitrary sum of money is set aside for maintenance and that this cannot be exceeded even though to achieve an acceptable standard would involve a greater expenditure. Thus the standard is really determined by the amount of money allocated rather than as a result of assessing the benefits obtained from maintaining the building to a particular state. A more functional definition proposed by White[11] is that 'maintenance is synonymous with controlling the condition of a building so that its pattern lies within specified regions'. The word 'control' suggests a positive activity which is planned so as to achieve a defined end result. The term 'specified regions' presumably has a meaning similar to 'an agreed standard' and would be determined in a similar way. An interesting aspect of the definition is that it envisages a range of acceptability with upper and lower limits between which the condition of the building must be maintained. Furthermore, this range of standards would be defined as much by the use of the building as by its physical state. So it follows that the concept of a well-maintained building as an entity in its own right is meaningless: a building is only well-maintained in terms of its current use and occupation. If this changes over time or through a change in ownership, the parameters of the acceptable level of maintenance will change in tandem. This point may be taken further: is it possible to have a building which is too well maintained? Certainly if the building remains in a better physical state than it need be in view of its use without extra maintenance costs, the point does not arise; but too great an expenditure on maintenance unnecessary to the use of the building may be constituted as waste. As Williams[12] has observed in this respect: 'many organisations adopting a planned maintenance strategy are beginning to suspect that their buildings are "over-maintained" to the detriment of funds for core business activity.' The essential point is that the acceptable level of maintenance is that which is targeted towards the expectations, use patterns, and resources of the controlling organisation. As such, maintenance should be regarded as a service provision directed towards the user rather than the attainment of a particular physical state of being for the building. It thus represents a balance rather than an absolute ideal.

1.5 Maintenance or improvement

The problem of distinguishing between maintenance and improvement is not solved satisfactorily by any of the aforementioned definitions. It is generally conceded that maintenance should include a reasonable element of improvement, e.g. the replacement of worn out components with up-to-date versions. As a result of technological developments replacements are rarely exactly the same as the original, being superior in some respects but perhaps inferior in others. For instance, a replacement boiler may have a better technical performance but a shorter life than the original. In such cases it is difficult to assess the notional element of improvement, especially where the modern component, as a result of improved manufacturing processes, is cheaper than the original. It is logical therefore to extend the meaning of maintenance to cover localised improvements of this sort. But where the intention is to increase the efficiency in the use of the building by adding facilities which were not previously present, the work should be classed as improvement. There are particular instances, such as in lease definitions of 'repair', where the boundary between maintenance and improvement must be defined with some care. In the case of leases, the presumption is that repairs should not constitute an improvement or betterment of the property funded by the lessees via the service charge for the benefit of the lessor on reversion. In this case the upgrading of an element to its modern equivalent, for example replacing single-glazed windows with modern double-glazed units would be acceptable, but upgrading the premises in a more general manner, for example building garages where there were none, would constitute an improvement.

The distinction between maintenance and improvement also assumes importance when viewed in the context of feedback to the designer. There is a tendency to believe that the level of expenditure on maintenance would be significantly reduced by better informed initial design. In order to test the validity of this belief, it is necessary to distinguish between work rendered necessary by technical design faults and that directed to improvements for purposes which could not have been foreseen by the designer and incorporated in the original scheme. This is dealt with in more detail in Chapter 10. The usefulness of the information fed back to the designer will therefore depend upon the extent to which it distinguishes:

(1) Work which may be regarded as normal in relation to the constructional materials and conditions of use. For instance, where the designer has made a conscious decision to use relatively short-lived materials for a particular purpose, the resulting high maintenance costs would be a normal consequence of that decision.
(2) Work resulting from design faults in relation to either
    • technical errors concerning inappropriate constructional details or materials, or,
    • errors of layout in relation to size, arrangement and juxtaposition of working spaces.
(3) Work made necessary by a demand for higher standards or a change in the pattern of use which could not have been foreseen at the time of the initial design.

Although technical design faults attract most attention, it is probable that layout faults prove very much more costly in the long run.

There is a further issue related to the maintenance/improvement debate, relating to conservation. What is constituted by one occupier as an improvement may be seen as a destruction of a valuable and irreplaceable feature by another. We thus have the concept of 'improvement' in a negative reference frame whereby older forms of construction may be valued because of their historicity and uniqueness, and are thus worthy of special attempts at preservation. Maintenance in this context may be taken literally, as an alternative to 'improvement' rather than as an adjunct. This aspect is covered in greater detail in Chapter 11.

Within the broad term 'maintenance' there are a number of more specific terminologies which will be examined next.

1.6 Planned preventive maintenance

BS 3811 defines the different types of maintenance in the following ways:

(1) *Planned maintenance.* Maintenance organised and carried out with forethought, control and the use of records to a predetermined plan.
(2) *Preventive maintenance.* Maintenance carried out at predetermined intervals or to other prescribed criteria and intended to reduce the likelihood of an item not meeting an acceptable condition.
(3) *Running maintenance.* Maintenance which can be carried out whilst an item is in service.

Preventive maintenance is normally planned and hence the term planned preventive maintenance. It is a concept which is probably more applicable to plant and equipment which is subject to mechanical wear but there are certain building elements which justify this treatment. In order to introduce such a system it is first necessary to produce an inventory of every building, area, service, etc., which has to be maintained. Then it is necessary to determine which items should be included in the planned preventive programme and the frequencies at which they will require attention, e.g. weekly, monthly, quarterly or annually. The selection would be based on the consequences of failure in regard to such factors as safety and productivity, e.g. fire doors would obviously be included, and the frequency on an analysis of past records. Finally works orders are prepared for the various tasks and an appropriate bring-forward system devised. Performance should be monitored continuously to check that the work is being carried out in accordance with the programme and that the costs are commensurate with the benefits, e.g. reduction in emergency work and user requests.

Bushell[13] suggests that planned preventive maintenance is worthwhile if:

- it is cost effective
- it is wanted to meet statutory or other legal requirements
- it meets a client need from an operating point of view
- it will reduce the incidence of running maintenance necessitating requisitions for work from the user

- there is a predominant incidence of work for the craftsman rather than pure inspection.

However, the point made earlier by Williams[12] should again be noted. A preventive maintenance system which carries out unnecessary work, or which maintains to an unnecessarily high level, is wasteful. This can come about particularly when repair needs are anticipated too far in advance of their actual failure. This may happen because an inspecting surveyor, who may lack the necessary knowledge of deterioration processes, assumes too high a required functional standard, or adopts an overcautious approach in order to minimise professional liability.

The key determinants in preventive programmes are twofold:

(1) *Servicing.* Scheduled work at regular defined intervals to prolong life and prevent breakdowns
(2) *Repairs and replacements.* Planned programmes based on observed condition, to repair or replace components immediately prior to anticipated failure. The concept of 'just-in-time' repairs is appropriate here.

Additionally one could add that the very act of planning, involving as it does an analysis of past performance in order to predict the future, leads to a more enlightened approach to the management of maintenance operations.

1.7 Asset planning and facilities management

In the case of large property-owning organisations it is essential to adopt a systematic approach to the overall management of building assets. In recent years this has come about through the growth of the facilities management profession, which may include building maintenance management within its remit but also entails a wider, if shallower, agenda relating to the holistic management of an organisation's total assets, which may include vehicles and other equipment in addition to the fixed assets. Facilities management can itself be seen as part of a wider process of asset management which deals with the sourcing and use of an organisation's entire assets.

In the context of building assets, the whole process of asset management should start at the design stage of each new building with a detailed analysis of the life cycle costs so as to achieve the right balance between initial construction costs and the subsequent maintenance and running costs. The building owner is thus made aware at an early stage of the amount that should be set aside for maintaining the building to an acceptable standard. It should be part of the designer's brief to provide a maintenance manual outlining amongst other things the main elemental life cycles and the future maintenance and energy requirements. Similar life cycle cost analyses should be carried out at regular intervals during the occupation period to take into account changes in the condition and use of the building(s). By this means a strategic asset plan can be drawn up providing for the long-term allocation of resources to the related activities of maintenance, refurbishment, new build and replacement, to reflect the changing building requirements of the organisation.

The following represents a systematic approach to asset planning.

Compile a detailed property database

The following is the sort of basic data needed:

- location
- age
- function
- number of storeys
- construction
- floor area
- element areas
- services
- space usage
- remaining life
- occupation costs
- replacement value
- site value
- constraints, e.g. listed building
- external works.

In the case of new buildings the information may be obtained from 'as built' drawings, bills of quantities suitably amended to allow for variations, or a maintenance manual. For older buildings in the absence of suitable records the necessary data would have to be compiled from inspection reports or special surveys. A format for this is given in an Appendix, Standard Maintenance Descriptions.

Determine the condition of the building(s)

For this purpose a condition survey would have to be carried out. This comprises a limited inspection concentrating on those elements at greatest risk and most likely to fail. In order to standardise the approach a checklist of the items to be inspected should be prepared with a simple grading system (also given in the Appendix) to record the severity of defects. For large housing estates of a similar type of construction and age it would be sufficient to take a random sample. The object is to give a general picture of the state of the buildings, from which the approximate cost of remedial work can be estimated.

Analyse the usage and performance of building spaces

There is little point in carrying out full-scale maintenance on building spaces which are under-utilised or not utilised at all. If maintenance expenditure is to be justified it must be shown that there is a return of some sort, whether it be financial or merely in terms of comfort and convenience.

The current and possible alternative uses of the building spaces should be considered, taking into account their functional efficiency and the effects of new legislation, new technology and changes in user requirements. In commercial and industrial buildings where requirements are subject to fairly rapid change the regular updating of the premises may be seen as an alternative to maintaining the original facilities. This approach has been described as 'planned obsolescence' and consists of a continuing programme of refurbishment to keep abreast of market trends. It is particularly appropriate to shopping areas where it is necessary to keep ahead of competitors in meeting changes in the shopping pattern of consumers and in retailing methods. Similar periodic assessments of standards are necessary for housing estates to provide for changes in family structure, social and work patterns and tenant preferences. The aim is to consider not only efficiency in satisfying present demands but also changes that will be needed to satisfy future demands.

Apply life cycle cost techniques to optimise all resource implications

This involves quantifying the costs and benefits of alternative strategies to meet the long-term needs of the organisation. For example, in the case of alterations to the internal layout of a commercial or industrial building the benefits to include in the equation may cover such matters as reductions in staff costs, lower running costs, more economic use of production machinery, improved public image, etc. Similar considerations would apply to repair/renew decisions in relation to the expected remaining life cycle of the building.

Formulate an investment programme itemising the expenditure requirements of the various activities and stating how the money is to be raised

The financial plan should itemise all the short- and long-term expenditures required for the proper upkeep and improvement of the buildings so that adequate provision can be made for the necessary funds. This would remove one of the objections that is sometimes made about repairs funds, that the expenditure is too uncertain. It would also be of great assistance in the case of buildings in multiple occupation where a service charge is levied on tenants and owners for the maintenance of the common parts and facilities. It would enable tenants to know well in advance what commitments they have for future maintenance costs and allow expenditure on large-scale repairs to be spread over a number of years.

Prepare an integrated action plan

This should set out the timing and resource needs of the operations necessary to ensure that the maintenance policy is part of and complementary to the total programme of maintenance, refurbishment, new build and replacement. It should include such matters as standards of upkeep, priorities, criteria for judging levels of maintenance expenditure, the organisation and decision-taking bases for the maintenance function and the degree of control required through the management information system.

1.8 Maintenance policy

Within the context of overall asset management, the maintenance policy should define the environment and ground rules for the delivery of the maintenance management service. BS 3811 defines maintenance policy as a strategy within which decisions on maintenance are taken. Alternatively it may be defined as the ground rules for the allocation of resources (workers, materials and money) between the alternative types of maintenance action that are available to management. In order to make a rational allocation of resources the benefits of those actions to the organisation as a whole must be identified and related to the costs involved. It is necessary, therefore, to consider the question of policy under the following headings.

Objectives

What does maintenance have to achieve? This should be viewed in the context of the organisation's overall building needs. Maintenance is an important part of the terotechnology approach which has been defined as a combination of management, financial, engineering and other practices applied to physical assets in pursuit of economic life cycle costs. It requires all departments in an organisation to co-operate in ensuring that the assets of the organisation are planned, provided, maintained, operated and disposed of at the lowest total cost to the organisation. Life cycle costing, or costs-in-use, is a technique which is usually thought of in terms of initial design decisions but it is equally useful for appraising expenditure on maintenance, alterations and improvements during the life of a building. It is explained in more detail below.

Benefits

What is to be gained? The benefits may be either short-term or long-term and may be classified as financial, technical or human. The financial benefits spring from a more effective use of the building and are reflected in higher productivity, less wastage of materials, improved sales figures, etc. The technical factors are related to the preservation of the physical characteristics of the building and its services and are reflected in fewer breakdowns with a reduction in downtime and fewer calls for emergency repairs, less accidents, lower future maintenance costs, etc. The human factors are related to the psychological effect of the condition of the building on the user and are reflected in such things as a lower rate of staff turnover with reduced recruiting and training costs, better customer relations and an improved public image. Clearly some of these benefits are difficult to quantify but some attempt should be made to express them in money terms so that the analysis can be as complete as possible. In any case, they should not be ignored simply because a book value cannot be put on them.

Policies

How shall we proceed? This involves laying down operational and cost objectives for the maintenance department starting with the identification of maintenance tasks, the standards to be achieved and the limits of cost. This will lead to policies concerning the proper balance between preventive and corrective maintenance, how far work should be programmed rather than relying on user requests, the priority to be accorded to different types of work, whether the work is better carried out by direct labour or contract and, where the properties are dispersed over a wide area, the extent to which decision taking should be decentralised. These policies will determine the structure of the maintenance organisation and the roles and duties of the supervisory staff.

A structure for a maintenance policy is given below[14].

The policy statement
- maintenance policy in relation to overall management policy in the organisation
- legal considerations and responsibilities
  lease terms, repair covenants, restrictions on access, applicable statutory standards
- environmental policy
  waste, noise, dust generation; materials sourcing, energy management (see Chapter 11)

The policy as it relates to building users
- standards of maintenance
- health and safety
- security and access
- work in occupied premises

Organisation

- management staff
  number, and split (if any) between those in-house and consultants
- preventive maintenance and the planning cycle
  proportion and type of preventive maintenance, timescales for inspections/ servicing
- identifying maintenance needs
  defect reporting, inspection types and cycle; prioritising maintenance needs
- routine servicing requirements and cycles

Procurement

- use of contractors and/or direct labour – proportion and types of work
- approved contractor list policy – contractor selection and vetting
- contracts: types, costs, timescales, levels of service
- emergency and disaster routines

Monitoring and feedback

- reporting to the management board
- quality management routines
- benchmarking

1.9 Life cycle costs

Life cycle costs (LCC) are the total costs of owning and using an asset over its predicted life span. Life cycle costing is a technique used to ascertain a suitable balance between capital expenditure on initial provision, and costs incurred as a consequence of use; for example energy consumption.

$$LCC = Ic + (Mc + Ec + Cc + Oc) + (Vc) - Rv$$

where

Ic is initial cost
Mc is maintenance cost
Ec is energy cost
Cc is cleaning cost
Oc is overhead and management cost
Vc is utilisation cost
Rv is resale value

The basic concept is that decisions on the design and acquisition of durable assets should take into account the long-term financial consequences and should not be based solely on initial costs. The LCC method can be used at all stages in the life cycle of a building from inception to eventual sale or demolition:

(1) At the inception stage it may be used to determine the most economic way of meeting a need for additional building space. The options to be compared may include:
   ● rearrangement of the internal spaces within an existing building
   ● building an extension
   ● gradual redevelopment of the existing site
   ● development of a new site
   ● purchase or lease of another building.
(2) During the early design stage of a new building to assist in developing the most economic plan shape, structural form and internal layout. It should be borne in mind that the earlier the LCC technique is applied, the greater the possible savings and the lower the committed costs. Thus the client's requirements should be stated initially in very broad terms so as not to limit the choice of options.
(3) During the detail design stage to identify the design features, components and finishings that have the lowest total costs. This stage would culminate in the preparation of a life cycle cost plan which should be incorporated in the maintenance manual for the guidance of the property manager.
(4) During the occupation of the building to assist in formulating planned maintenance and renewal policies. It also provides a means of identifying high cost areas and evaluating changes that will reduce these costs.

The costs may be broadly divided into capital and revenue, a distinction which is usually necessary for accounting and taxation purposes. Unfortunately the levels of these two types of expenditure are generally determined by different decision processes and are subject to different control systems and separately funded. Because of this the significance of total costs is obscured. For LCC to be effective it is essential that decisions on initial capital costs should be tempered by a knowledge of the consequential running costs. One way to ensure that this is done is to require design proposals to be accompanied by an LCC plan.

Figure 1.7 shows the main life cycle costs and the approximate percentages of initial cost for the various items of revenue expenditure. The percentages are based on an analysis of the life cycle costs of an office block and as will be seen the total revenue costs amount to about 10% of the initial costs.

The approach to LCC should follow the pattern illustrated in Fig. 1.8 and itemised below:

(1) Establish basic objectives. This should be a clear statement of what the proposals are intended to achieve and may range from the life cycle costs of a new town to the choice of roof coverings for an individual building. For the property manager it could include an evaluation of alterations to

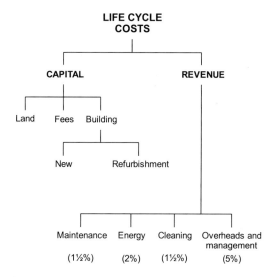

Fig. 1.7 Life cycle costs.

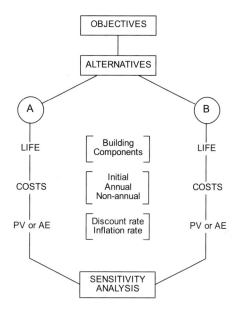

Fig. 1.8 LCC method.

an existing building or the financial consequences of renewing rather than repairing a major piece of equipment.

(2) Formulate alternative means for achieving the stated objectives.

(3) Decide upon a finite planning horizon which is applicable to all the alternatives.

(4) Identify all the costs and revenues which are directly relevant to the comparison of the alternatives.
(5) Adjust the costs to a common time period by converting to present values or annual equivalents.
(6) Carry out a sensitivity analysis to assess the effect of errors in predicting building and component life and different rates of interest.

Flanagan and Norman[15] identify three distinct but closely related aspects of LCC, which are:

(1) Life cycle cost planning (LCCP), which focuses on the planning of future costs as well as initial costs.
(2) Life cycle cost analysis (LCCA), which involves the systematic collection of running cost data on completed buildings and linking this with the physical, qualitative and performance characteristics of those buildings.
(3) Life cycle cost management (LCCM), which covers the application of the techniques to existing buildings and systems and has the following aims:
    ● to monitor and explain differences between LCCP projections and actual performance
    ● to improve the efficiency of the building through more effective utilisation
    ● to provide information on asset lives and reliability factors for accounting purposes
    ● to aid in the development of an appropriate maintenance policy for the building
    ● to provide taxation advice on building related items.

Thus LCCM forms an important part of property or asset management.

1.10 Maintenance generators

The agencies which act upon the building and erode the initial standards include:

(1) *Climatic conditions* which vary in severity according to the location and orientation of the building and which have the greatest effect on the external elements.
(2) *User activities* including both human and mechanical agencies and authorised and unauthorised usage. For this purpose a burglar may be considered to be a user of the building, albeit an illegal and unwelcome one.
(3) *Changing standards and tastes* which, while not worsening the existing condition, may create a demand for work to be carried out more frequently

than functionally necessary; for example, repainting for the sole purpose of changing the colour scheme.

In all these examples, the primary cause may be regarded as either 'normal' or 'abnormal' according to expectations based on past experience. Also, the work required may relate to preserving either the 'functional' or the 'aesthetic' properties of the facility.

The extent to which these agencies cause deterioration and thus create a need for remedial treatment will depend upon:

- the adequacy of the design and the suitability of the materials specified
- the standard of workmanship in the initial construction and subsequent maintenance
- the extent to which the designer has anticipated future needs.

Over a period of time the gap between the standards demanded by the user and those provided by the building is likely to widen as shown in Fig. 1.9. The diagram is an abstract representation of the building as a whole and quite clearly is not applicable to each and every element, e.g. the requirements for the foundations are unlikely to change and, in the absence of a design error, should last the life of the building without attention. Also, individual elements will vary considerably with regard to the rate at which changes take place and the magnitude of the changes.

Another factor which is difficult to incorporate in a model of this type is the interdependence of elements and the change in the pattern of deterioration caused by the failure of other elements. Thus an apparently minor defect may have consequences, as illustrated in Fig. 1.10.

Admittedly this particular chain of events could have been anticipated by the designer, and a layer of felt provided underneath the tiles to convey safely to the gutter any rainwater that might penetrate the covering. However, while this second line of defence would operate satisfactorily for a time, excessive delay in replacing the tile could lead to consequential damage which, although not

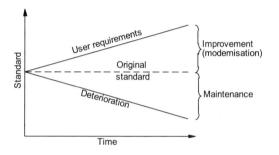

Fig. 1.9   The passage of time related to maintenance requirements.

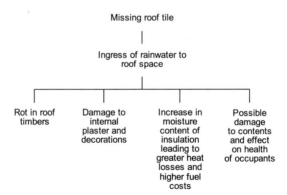

Fig. 1.10   Consequences of defective maintenance.

capable of exact prior evaluation, would clearly be more costly to remedy than the timely replacement of the tile. And, of course, a similar chain of events could be triggered off by other quite different causes, e.g. excessive condensation.

The operation of factors generating the need for maintenance is dealt with more rigorously in Chapter 10.

1.11 Timing of maintenance operations

The work necessary to combat progressive deterioration takes the following forms:

- Patching, involving the more or less regular replacement of small parts or areas.
- Replacement of whole elements of components because they:
  - o   are functionally unsatisfactory;
  - o   incur high maintenance or running costs;
  - o   are aesthetically unacceptable.
- Preservation of protective coatings either for the purpose of extending the life of the protected material or to maintain appearance.
- Cleaning which, although often regarded as a separate activity, has important maintenance implications in arresting deterioration and preserving appearance.

In many cases, the precise outcome of the reaction between the exposure and use conditions and the resisting properties of the building elements cannot be known with certainty. However, some idea as to the probability of failure may be gained by grouping building elements as follows:

(1)   Those which if properly designed should last the life of the building without requiring attention. Foundations fall into this category, any work neces-

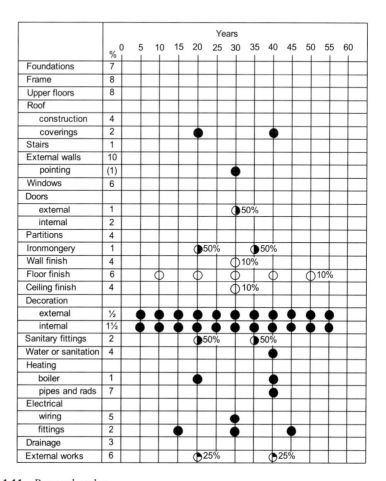

Fig. 1.11 Renewal cycles.

sary such as underpinning being attributable to faulty design or an unforeseen change in ground conditions.

(2) Those whose life can be prolonged by the replacement of small parts at more or less regular intervals. This applies to such elements as roof tiling, although a time would come when it would be more economic and functionally more satisfactory to renew the whole of the coverings.

(3) Those components which are subject to wear through either human or mechanical agencies. Thus the life of floor finishes is related to the density and type of traffic.

(4) Those which are prone to obsolescence as a result of technological advances or merely changing fashion. This applies to all types of surface finishes and fittings such as sanitary appliances. One finds, for example, that the appearance of glazed brick dados in older schools is now considered too 'institutional'; in spite of their excellent maintenance qualities, they are often covered with other finishes which have inferior maintenance

characteristics. It is difficult to apply a time scale to changes in taste but it is reasonable to suppose that the rate of change will increase and there will be an increasing tendency to reject components which, while still functionally adequate, do not measure up to the accepted 'norms' of the period.

(5) Those which are exposed to the weather and will in the course of time fail or become disfigured to such an extent that repairs or extensive cleaning will be necessary. This category includes roof coverings and external wall claddings and facings of all types.

(6) Protective coatings: these require special consideration in that, although the coating may have some aesthetic value, the primary reason for renewal is to extend the life of the protected element. This type of work forms a high proportion of the total maintenance expenditure and methods of optimising renewal cycles will be considered later.

1.12 Maintenance cycles

Assuming normal conditions of exposure and use the periodicity of major renewals might follow the pattern indicated in Fig. 1.11. For this purpose a building life of 60 years has been taken and the renewal periods estimated in multiples of five years. In addition to the major renewals and replacements there will be a large number of smaller items which are of uncertain timing but which in total tend to remain fairly constant from year to year.

If it is assumed that these routine maintenance costs amount to 0.5% of the initial cost each year, i.e., 2.5% of initial cost per quinquennium, the overall cash flow over the life of the building will be as shown in Table 1.1. This has been

Table 1.1 Cash flow at current prices as percentage of initial cost.

Year	Percentage initial cost per quinquennium	Percentage initial cost cumulated
5	4.5	4.5
10	5.1	9.6
15	6.5	16.1
20	11.1	27.2
25	4.5	31.7
30	14.4	46.1
35	4.5	50.6
40	20.6	71.2
45	6.5	77.7
50	5.1	82.8
55	4.5	87.3
60	2.5	89.8

derived from the percentages and renewal frequencies shown in Fig. 1.11. The percentages shown in the second column of Fig. 1.11 are the percentages of the total initial cost attributable to the individual elements for the building on which the analysis is based. Thus the maintenance cost for the first quinquennium is made up of 0.5 (external decoration) + 1.5 (internal decoration) + 2.5 (routine) = 4.5%. This is given purely as an example of a method which can be used to assess maintenance costs over the life of the building. Clearly different buildings will have different elemental cost patterns and renewal periods will vary according to design, location and use.

In this example the maintenance costs have been expressed as a percentage of the initial cost of the building but, given the initial cost, they could be expressed in monetary terms either for the building as a whole or per 100 m² of floor area.

No attempt has been made to allow for the effects of inflation on future maintenance costs, on the assumption that the funds which bear the expenditure will inflate at the same rate. However, it is unlikely that maintenance costs and new construction costs will continue to bear a constant relationship over a prolonged period of time. Labour forms a higher proportion of the cost of maintenance than of new construction and as wage rates are currently increasing at a faster rate than the prices of most materials it is to be expected that maintenance will become more expensive in relation to new construction.

Also, no account has been taken of the time scale of the payments, i.e., that future disbursements can be met by investing smaller sums now at interest. For instance, £100 invested now at 10% will be worth £110 after one year or, conversely, £110 at the end of the year is worth only £100 now. The present value (PV) of future sums can be calculated from the following formula:

$$PV \text{ of } £1 = 1 / (1 + i)^n$$

where i = rate of interest + 100 and n = number of years.

In practice such calculations are usually done by taking the appropriate factor from a 'present value of £1' table and multiplying the future sum by this factor to obtain the discounted amount. The following is an extract from such a table:

Present value of £1

Years	Interest		
	6%	9%	12%
5	0.7473	0.6499	0.5674
10	0.5584	0.4224	0.3220
20	0.3118	0.1784	0.1037
30	0.1741	0.0754	0.0334
60	0.0303	0.0057	0.0011

Thus if the cost of replacing a component in ten years' time is £100 and the rate of interest is assumed to be 9%, the PV of the replacement is $0.4224 \times £100 = £42.24$.

Where regular annual sums are expended over a period of years the present value of the recurring sums may be obtained from the following formula.

$$\text{PV of £1 per annum} = (1+i)^n - 1 / i(1+i)^n$$

Again, it is more convenient to use factors from standard tables, e.g.

Present value of £1 per annum

Years	Interest		
	6%	9%	12%
5	4.212	3.890	3.605
10	7.360	6.418	5.650
20	11.470	9.129	7.469
30	13.765	10.274	8.055
60	16.161	11.048	8.324

Thus if the annual cleaning costs over the 60-year life of a building are £500 the present value, assuming a 9% rate of interest, is $11.048 \times £500 = £5524$. This table is also known for valuation purposes as 'years purchase – single rate' and can be used for finding the annual equivalent of an initial capital sum. For instance, if in the previous example the object had been to find the annual amounts which an investment of £5524 would yield over a period of 60 years at 9% interest, the figure would be $£5524 / 11.048 = £500$. At the end of the period both the capital and interest would be expended unless an additional allowance had been made for a sinking fund.

Applying discounting factors to the cash flows given in Table 1.1, the pattern of expenditure shown in Table 1.2 is obtained.

The cumulative expenditure is shown in graphical form in Fig. 1.12 and it will be seen that as the interest rate increases the future maintenance costs appear less and less significant. This effect is even more pronounced when taxation is taken into account. As a general rule there is no tax relief for the initial capital cost of a building, but maintenance and running costs may be set against taxable profits. Thus, assuming a rate of 50% for corporation tax, the true cost to the firm of maintenance is only 50% of the actual cost. As opposed to this, maintenance and alterations are subject to value added tax, but if the building owner is a 'taxable person' the tax is recoverable.

Table 1.2   Discounted cash flow as percentage of initial cost.

Year	Percentage initial cost discounted at 6%		Percentage initial cost discounted at 9%		Percentage initial cost discounted at 12%	
		Cumulative		Cumulative		Cumulative
5	3.38	3.38	2.93	2.93	2.57	2.57
10	2.86	6.24	2.14	5.07	1.63	4.20
15	2.73	8.97	1.76	6.83	1.17	5.37
20	3.44	12.41	2.00	8.83	1.11	6.48
25	1.04	13.45	0.54	9.37	0.27	6.75
30	2.45	15.90	1.15	10.52	0.43	7.18
35	0.59	16.49	0.23	10.75	0.09	7.27
40	2.06	18.55	0.62	11.37	0.21	7.48
45	0.46	19.01	0.13	11.50	0.07	7.55
50	0.26	19.27	0.05	11.55	0.02	7.57
55	0.18	19.45	0.05	11.60	0.01	7.58
60	0.08	19.53	0.02	11.62	—	7.58

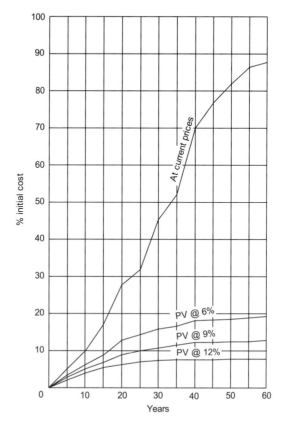

Fig. 1.12   Maintenance costs.

1.13 Maintenance profiles

The pattern of maintenance costs over the life of a building is known as the 'maintenance profile' and is usually presented in the form of a histogram. The example in Fig. 1.13 is based on the maintenance costs given in Table 1.1. The peaks of expenditure are attributable to major renewals and replacements which could be deferred by incurring higher running repair costs and perhaps some loss of amenity. Where the estate consists of a number of buildings some adjustment could be made to the frequency of renewals and cyclic work for the individual buildings so as to achieve a more uniform total workload from year to year. This would not only give a more acceptable cash flow but would be essential where it is the policy to do as much work as possible by direct labour.

However, it is somewhat unrealistic to relate predicted costs to particular years in the life of a building. Costs are made up of a large number of small items of uncertain timing together with a limited number of major items of renewal or replacement whose timing will depend, for the most part, on factors unconnected with the physical condition of the building. The uncertainties inherent in the timing of these major items can be reflected by assessing the probabilities of the work being necessary at different times. The probability scale runs from zero to unity, the bottom end of the scale representing absolute impossibility and the top end absolute certainty. In the context of maintenance few events are so clear-cut as this and in most cases the probability that the event will occur lies somewhere between these extremes. For instance, if there is a 50:50 chance that a particular component will have to be replaced in five years time, the probability would be expressed as 0.5, or in mathematical terms $p = 0.5$.

Applying probability theory to the renewal times for the felt roof covering, Fig. 1.14 indicates renewal in years 20 and 40 but for a great variety of reasons failure may occur earlier or later than anticipated. On the basis of past experience one might assess the probabilities as in Table 1.3. Years 20 and 40 are rated as the most probable times for the first and second renewals but, as the second renewal is dependent on the timing of the first, the probability that this will occur

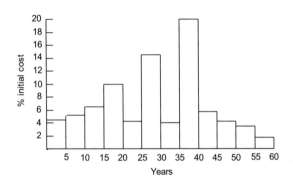

Fig. 1.13   Maintenance profile.

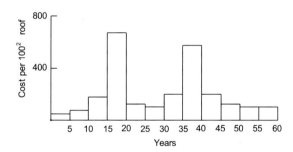

Fig. 1.14　Maintenance profile of roof.

Table 1.3　Probable incidence of renewal times.

Year	First renewal	Second renewal	Total probability of renewal
5	—	—	—
10	—	—	—
15	0.1	—	0.1
20	0.8	—	0.8
25	0.1	—	0.1
30	—	0.01	0.01
35	—	0.16	0.16
40	—	0.66	0.66
45	—	0.16	0.16
50	—	0.01	0.01
55	—	—	—
60	—	—	—

in the fortieth year is slightly less than the probability of the first renewal occurring in year 20. It has been assumed that as the life of the building is 60 years, renewal after the fiftieth year would not be contemplated.

The total maintenance cost for the roof will be made up of:

- renewal of the coverings
- patching of the coverings between renewals
- repairs to constructional members and eaves and rainwater goods

as shown in Table 1.4.

The cost of patching has been related to the renewal periods but a constant sum has been assumed for repairs to structural members and rainwater gutters and pipes. This will only hold good if patching is carried out promptly so as to maintain the coverings in a waterproof condition. The maintenance profile of the roof is as shown in Fig. 1.14.

Table 1.4 Maintenance cost of roof.

Year	Renewal of coverings (assume cost of renewal including all labours £8 per m²)		Patching between renewals	Repairs to structure such as gutters, etc.	Total
		£ per 100m²	£ per 100m²	£ per 100m²	£ per 100m²
5	—		20	20	40
10	—		40	20	60
15	0.1 × 800	80	60	20	160
20	0.8 × 800	640	20	20	680
25	0.1 × 800	80	40	20	140
30	0.01 × 800	8	60	20	88
35	0.16 × 800	128	60	20	208
40	0.66 × 800	528	40	20	588
45	0.16 × 800	128	60	20	208
50	0.01 × 800	8	80	20	108
55	—		80	20	100
60	—		80	20	100

Similar maintenance profiles can be produced for other building elements to illustrate the timing and magnitude of the costs involved. Care should be exercised when applying costs and renewal cycles obtained from other buildings. Although the design and materials specification may be similar, the periodicity of repairing cycles may vary greatly according to climatic conditions and user activities. Even with the same building a change of use or an intensification of the existing use can have a profound effect on the maintenance profile. In particular, finishings and fittings may have a very much shorter life if subjected to more rigorous use conditions than those envisaged by the designer.

1.14 Optimising renewal cycles

The assessment of renewal cycles is based on knowledge of the rates of deterioration of similar buildings under similar conditions of exposure and use. Although the subjective element cannot be ruled out altogether it is possible to introduce a degree of objectivity into the assessment. One way in which this can be done is to prepare an anticipated state matrix for the element under consideration by assessing its probable condition at regular intervals throughout its life and estimating the cost of the remedial work necessary at those times. For instance, for the felt roof covering considered earlier the probable states and remedial costs may be as shown in Table 1.5.

Table 1.5 Renewal cycle of roof covering.

Anticipated state		Remedial work	Cost per m²
1	No visible defects	Nil	—
2	Small isolated blisters	Nil	—
3	Large blisters and slight cracking causing minor localised leaks	Patching	£12
4	Extensive cracking and deterioration causing widespread leaks	Renew	£8

Year	5	10	15	20	25
State	Percentage of roof coverings in each state				
1	80	60	30	15	10
2	19	25	35	25	20
3	1	10	25	30	20
4	—	5	10	30	50

Maintenance cost
per m² (patching) $0.01 \times 12 = £0.12$ $0.15 \times 12 = £1.80$ $0.35 \times 12 = £4.20$ $0.60 \times 12 = £7.20$ $0.70 \times 12 = £8.40*$
*Cost of patching exceeds renewal cost.

The analysis confirms the earlier presumption that the most economic time to renew the coverings would be about the twentieth year. It is probable that the price per square metre for patching would be reduced slightly in later years because of the larger areas requiring attention but nevertheless, with 30% of the coverings reduced to state 4, it is unlikely that the roof would be able to perform its primary functions satisfactorily. It has been assumed that patching will be carried out reasonably promptly after leaks are reported; otherwise it would be necessary to make some allowance for the probable deterioration of the roof timbers and the effect on the user of the building.

A similar form of analysis can be used to optimise the cycles for external painting. For this it is necessary to predict the probable treatments required at intervals of one year up to, say, ten years, and then identify the period which gives the lowest annual costs. The deterioration pattern will depend upon such factors as the quality of the paint, the thoroughness of the preparatory work, the severity of the exposure to atmospheric pollution and ultraviolet light and other matters peculiar to the particular building and the orientation of the elevation. For the purposes of this example it is assumed that the treatments required at the end of each period are as follows:

Year	
1	Wash down
2	Wash down and one coat of paint
3	Wash down and two coats of paint
4	Wash down, burn off 2.5% and prime and two coats paint
5	Wash down, burn off 5% and prime and two coats paint
6	Wash down, burn off 10% and prime and two coats paint
7	Wash down, burn off 20%, renew 2.5% woodwork, prime and two coats paint
8	Wash down, burn off 40%, renew 5% woodwork, prime and two coats paint
9	Wash down, burn off 60%, renew 10% woodwork, prime and two coats paint
10	Burn off 100%, renew 20% woodwork, prime and two coats paint

The costs of the assumed treatments can then be calculated as shown in Table 1.6.

On the basis of the assumptions made in this example the optimum period for external repainting would be six years. Washing down gives the same annual cost but even if this treatment were repeated each year it would be necessary after two or three years to apply one or two coats of paint to make good surface wear and defects as a result of movement of the substrate. Also, where appearance is of importance, it will be necessary to form a subjective assessment as to whether or not the state of the paint work at the end of the economic period will be considered acceptable. The difference in cost between repainting at the cal-

Table 1.6 Percentage of windows requiring treatment (based on 100 m² of softwood casement windows, measured overall).

Predicted treatments	Year									
	1	2	3	4	5	6	7	8	9	10
Washdown (£0.60 m²)	100	100	100	97.5	95	90	80	60	40	—
One coat (£1.20 m²)		100								
Two coats (£2.40 m²)			100	100	100	100	100	100	100	100
Burnoff (£4.50 m²)				2.5	5	10	20	40	60	100
Prime & stop (£2.10 m²)				2.5	5	10	20	40	60	100
Renew woodwork (£60 m²)							2.5	5	10	20
Cost of treatment (£)	60	180	300	315	330	360	570	840	1260	2100
Cost per year (£)	60	90	100	79	66	60	81	105	140	210

culated economic periods and at those judged necessary to maintain a satisfactory appearance will represent the value which the building owner or occupier attaches to appearance. Also the future costs could be discounted to allow for the time value of money.

1.15 Repair/replace decision

Appraisal methods

There are several methods of analysing the economic consequences of alternative decisions relating to the repair or replacement of building elements and components. The more common methods are as follows.

Payback period
The payback period is the length of time taken for the returns from an investment to equal the initial outlay. Where the proceeds are the same each year the payback period is obtained by dividing the initial outlay by the annual proceeds. Thus if roof insulation will cost £100 and the annual saving on fuel is estimated as £20 the payback period is five years. This can then be compared with the payback periods for other types of insulation; e.g. cavity-wall filling or double glazing to determine which type offers the best value. The method suffers from two main disadvantages:

- it does not take into account the time value of money,
- it ignores any benefits that may accrue after the end of the payback period.

It is, however, very simple and easy to understand.

Net present value (NPV)
This method is also known as the present worth method and involves discounting all future cash flows to a common base year. The formulas given earlier in this chapter for PV of £1 and PV of £1 per annum may be used for this purpose although it is probably easier to look up the factors in the standard tables. It should be noted that the analysis is particularly sensitive to the rate of interest used and, therefore, this must be chosen with care.

For example the total PV of repainting a building every five years over an anticipated life of 30 years (last repainting in year 25) will vary as follows for different rates of interest. For the purpose of this example the repainting cost has been estimated to be £200 and no allowance made for inflation.

PV of £1

Year	6% factor	9% factor	12% factor
5	0.7473	0.6499	0.5674
10	0.5584	0.4224	0.3220
15	0.4173	0.2745	0.1827
20	0.3118	0.1784	0.1037
25	0.2330	0.1160	0.0588
Total	2.2678 × £200 = £454	1.6412 × £200 = £328	1.2346 × £200 = £247

It will be noted that the higher the rate of interest the lower the PV of future costs and the less important maintenance costs appear to be in comparison with the initial costs of construction. Also, if the above analysis were being used to determine how much it would be worthwhile spending on finishes which would not require periodic repainting the decision would obviously depend upon the rate of interest used.

The method may also be used to produce a modified form of the payback method that takes into account the time value of money. Thus the example given in which the expenditure of £100 on insulation results in a £20 saving on fuel costs per annum could be analysed as follows:

Year	Saving (not adjusted for inflation)	PV of £1 factor for 9%	PV of unadjusted saving	Saving adjusted for 10% inflation per annum	PV of adjusted saving
1	20	0.917	18.34	20	18.34
2	20	0.842	16.84	22	18.52
3	20	0.772	15.44	24.2	18.68
4	20	0.708	14.16	26.6	18.83
5	20	0.650	13.00	29.3	19.05
6	20	0.596	11.92		
7	20	0.547	10.94		
			100.64		93.42
			7 years		about 5.5 years

Thus, if no account is taken of inflation, the payback period when considering the present values of the future savings is seven years. If a rate of inflation of 10% is assumed the payback period is reduced to about five and a half years. A larger increase in the price of fuel would reduce the period still further and strengthen the case for providing the insulation.

Annual equivalent (AE)

In this method the cash flows throughout the life of the asset are converted into an equivalent uniform annual cost. It will give the same ranking as the present value method but the presentation in the form of annual outgoings (or receipts) is probably more meaningful to the building owner. Where the cash flows are irregular it is easier to convert them all to their present value initially (except those which are regular annual payments and do not require converting) and then calculate the annual equivalent of the total present value. This is done by multiplying the total present value by the reciprocal of the PV of £1 per annum factor. This factor is known as the uniform series that £1 will buy and is given by the following formula.

$$\text{Uniform series that £1 will buy} = i(1+i)^n / (1+i)^n - 1$$

Again the rate of interest can have a significant effect on the ranking of alternative courses of action. Also difficulties can arise when comparing alternatives which have different lives. In such cases it is usual to base the analysis on a period of time which is the lowest common multiplier of the lives of the alternatives. Thus if a comparison is being made between a traditional brick building with an assumed life of 60 years and a timber-framed building with lightweight claddings and an assumed life of 30 years the analysis would be based on a 60-year period. This would necessitate including for the renewal of the shorter-lived building at the end of 30 years. Alternatively, if the analysis is based on the 30-year period an allowance should be made for the value of the traditional building at the end of that period otherwise the comparison is not being made on a common basis.

In some cases payments increase by a regular amount each year. The annual equivalent of such increases can be found using the following formula.

$$\text{Uniform gradient series} = g / i - ng / i \times (i / (1+i)^n - 1)$$

For example, assuming the maintenance costs for a building are £2000 in year 1 and increase by £50 per annum throughout its 60-year life, the annual equivalent of the increase at 10% interest will equal:

$$50 / 0.1 - (60 \times 50) / 0.1 \times 0.1 / ((1.1)^{60} - 1) = £490$$

This added to the regular component of £2000 gives a total annual equivalent cost of £2490.

Yield method

This is also referred to as the internal rate of return method or in some textbooks as discounted cash flow (DCF) although strictly speaking DCF would also include the PV method. The main difference between this method and the previous two methods is that in the present value and annual equivalent methods a fixed interest rate has to be assumed whereas the object of the yield method is to find the interest rate which will equate the total present value of all the future cash flows, both negative and positive, to the initial outlay.

For example, it has been estimated that the provision of certain mechanical handling equipment in a warehouse would reduce labour costs by £6000 per annum. The cost of installing the equipment is £20 000 and will require additional running costs of £1000 per annum. What is the rate of return from the investment assuming that the warehouse will only be used for a further five years?

Year	Receipts (£)	Payments (£)	Net cash flow (£)
0	—	−20 000	−20 000
1	+6 000	−1 000	+5 000
2	+6 000	−1 000	+5 000
3	+6 000	−1 000	+5 000
4	+6 000	−1 000	+5 000
5	+6 000	−1 000	+5 000
Total	+30 000	−25 000	+5 000

The first thing to do is to find the net cash flows over the period as shown in the preceding table. Then it is necessary to find the interest rate that will discount the future cash flows to give a total PV that is equal to the initial outlay of £20 000. This is found by trial and error although the following formula will give an approximate value for short periods:

$$\text{Interest rate} = \frac{\text{Total proceeds} - \text{Initial cost} \times 100}{\text{Period in years} \times \text{initial cost}}$$

(multiply by 2 for longer periods over 10 years)

The formula gives an approximate rate of 5% and this is used as a starter. As the total PV so obtained is greater than the initial outlay, i.e., £21 648, the true interest rate must be higher than 5%. Therefore, a rate of 9% is tried and this time the total PV is less than the initial outlay, indicating that the required interest rate must lie between 5 and 9%. The actual rate is found by interpolation as follows:

$$\text{Yield (\%)} = 5 + \frac{21\,648 - 20\,000}{21\,648 - 19\,499} \times (9 - 5)$$

$$= 5 + (1648 / 2199) \times 4 = 8.00\%$$

Year	Net cash flow (NCF)	PV of £1 factor 5%	PV of NCF at 5%	PV of £1 factor 9%	PV of NCF at 9%
0	−20 000	1.0000	−20 000	1.0000	−20 000
1	+5 000	0.9524	4 762	0.9174	4 587
2	+5 000	0.9070	4 535	0.8417	4 209
3	+5 000	0.8636	4 319	0.7722	3 861
4	+5 000	0.8227	4 114	0.7084	3 542
5	+5 000	0.7835	3 918	0.6500	3 250
Total	+25 000		+21 648		+19 449

The calculation can be checked by discounting the cash flows using an interest rate of 8% when the total PV should equal £20 000 (approximately). Whether or not the investment is considered worthwhile will depend upon the rate of return that can be obtained from an alternative use of the capital.

Break-even analysis

This method is concerned with making comparisons between two alternatives where the cost of each alternative is affected by a single common variable. The comparison may be made either mathematically or graphically and involves finding the point at which the value of the two alternatives is the same.

For example, a choice is to be made between two components A and B in which A has a lower first cost but higher maintenance costs than B. This can be represented as in Fig. 1.15 which shows that if the period of use is less than 20 years A is more economic whilst if greater than 20 years B is to be preferred.

Decision trees

The decision tree attempts to account for future uncertainties by evaluating all possible outcomes using subjective probabilities and expected values. The expected value of an event is its probability of occurrence multiplied by its conditional value, the conditional value being the loss or gain if the event in question

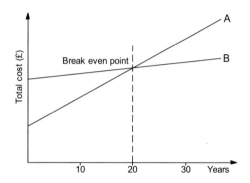

Fig. 1.15   Break-even point.

actually occurs. Thus if the consequential losses flowing from the breakdown of a machine are estimated to be £1000 and the probability of a breakdown occurring is 0.1, the expected value is £100. This gives some measure of the amount which it is worthwhile spending in order to avoid such a breakdown.

Simulation

This involves designing a model, usually utilising a computer, which closely resembles the operation being studied. Many variables are built into the model but no attempt made to estimate their precise values or frequencies of occurrence. Then the variables are systematically altered within their possible ranges in order to determine the effect of the different values on the working of the model and to identify the optimum solution.

Replacement

Generally replacement of a component will be considered when either

- it is functionally or aesthetically unsatisfactory, or
- the repair and running costs are excessive.

The choice is then between replacement with an identical component or with one of a different type which has a better performance. In such cases comparative costs-in-use analyses can be prepared to assist in deciding which alternative offers the best value for money. Often, however, a component is repairable and the decision as to whether to repair or replace will depend upon the repair costs and lives of the alternative courses of action. If it is assumed that the life of the repaired component is shorter than that of the replacement then it is quite simple to calculate the maximum economic amount to spend on the repair. For instance if the repair life is five years and a replacement would last ten years it would not be economic to spend more than 57% of the cost of replacement on the repair, assuming an interest rate of 6%.

This is calculated as follows:

Assume £x=repair cost, £y=replacement cost and interest rate=6%

£x	#	#	#	#	#					
Years					5					
Annual equivalent=$x/4.212$*										

£y	#	#	#	#	#	#	#	#	#	#
Years										
10										
Annual equivalent=$y/7.360$*										

*from PV of £1 per annum (year's purchase) table

For x and y to give equal value:

$$x/4.212 = y/7.360 \qquad x = 0.572y$$

Therefore, it is uneconomic if x, the repair cost, is more than 57.2% of y, the replacement cost.

For ease of reference this can be set down in tabular form as in Table 1.7. As will be seen, the higher the rate of interest the more worthwhile it is to take the shorter-term view and spend a higher proportion on repair, even though the life of the repaired component is shorter than a replacement.

Table 1.7   Maximum repair cost as percentage of replacement cost.

Repair life (years)	Replacement life (years)				
1	10	15	20	25	30
Interest 6%					
1	12.8	9.7	8.2	7.4	6.9
2	25.0	18.9	16.0	14.3	13.3
3	36.3	27.5	23.3	20.9	19.4
4	47.1	35.7	30.2	27.1	25.2
5	57.2	43.4	36.7	32.9	30.6
Interest 9%					
1	14.3	11.4	10.0	9.3	8.9
2	27.4	21.8	19.3	17.9	17.1
3	39.4	31.4	27.7	25.8	24.6
4	50.5	40.2	35.5	33.0	31.5
5	60.6	48.3	42.6	39.6	37.9

However, where the replacement is of a more advanced design than the original any ensuing benefits must be taken into account. This is particularly so where the replacement offers a saving in running costs. For example, if it is to be decided whether to repair an existing boiler at a cost of £200 or replace it with a more efficient type at a cost of £500 the 'total costs' of the two alternatives must be calculated. If it is assumed that the repaired boiler will have a life of ten years and consume £250 of fuel each year whereas the replacement will have a life of 15 years and consume only £200 of fuel each year, the total costs will be as follows:

Repair		
Annual equivalent of £200 over 10 years at 6% interest	$200/7.360 =$	27
Fuel costs		250
Total annual cost		£277

Replacement		
Annual equivalent of £500 over 15 years at 6% interest	$500/9.712 =$	51
Fuel costs		200
Total annual cost		£251

A replacement boiler would therefore represent the better value. This method of analysis may also be used when considering changing a boiler for one which uses a different type of fuel, although a forecast would have to be made of the future price levels of the fuels in order to identify the most advantageous long-term solution.

Similarly, the high cost of electricity may make it worthwhile to replace tungsten lamps with fluorescent tubes. The amount which it is worthwhile spending on effecting the change can be determined by calculating the savings which would be achieved over an appropriate period. For example, assume that a space is lit at present by means of one hundred 150 W tungsten lamps and it is required to find out how much it would be worthwhile spending on replacing them with a similar number of 40 W fluorescent tubes. It is estimated that the lighting will be used for about 1000 hours per year and that the comparative data are as follows:

	Life hours	Bulk replacement period	Replacement costs per lamp/tube £	Electricity consumption W
Tungsten lamps	1000	Annually	0.50	150
Fluorescent tubes	7000	Every 7 years	2.50	50 (extra 10 W consumed in control gear)

Analysis based on assumed rate of interest of 9% throughout life of fittings

---

Tungsten lamps

Annual cost		
Bulk replacement	100 lamps @ £0.50 each	£50.00
Electricity	$(100 \times 50 \times 1000)/1000 = 15\,000$ units @ 4p per unit	£600.00
Total		£650.00
Factor for PV of £1 p.a. for 14 years @ 9%		×7.786
Total PV =		£5061.00

Fluorescent tubes

Annual cost		
Electricity	$(100 \times 50 \times 1000)/1000 = 5000$ units @ 4p per unit	£200.00
Factor for PV of £1 p.a. for 14 years @ 9%		×7.786
		£1557.00
Bulk replacement		
after 7 years	100 tubes @ £2.50 each	£250
Factor for PV of £1 in 7 years @ 9% × 0.5470		£137.00
Total PV =		£1694.00

Saving over 14 years = £5061−£1694 = £3367

(Cleaning costs assumed to be the same for both types of fittings)

Thus it would be worthwhile spending up to about £33 per lighting point on the change. If the lighting is used for more than the estimated 1000 hours per year or if there is an increase in the price of electricity the change becomes even more worthwhile.

Other possibilities that could be considered in order to reduce electricity consumption on lighting range from the simple one of ensuring that lights are switched off when not required to providing separate switching for different areas with separate task and ambient lighting systems. Also, one well-known department store found that substantial savings could be made by reducing the level of illumination in the sales areas from 1000 lux to 600 lux without any detrimental effect on sales.

Alterations

Decisions relating to alterations are based on similar considerations to those that apply to repair or replace decisions. In both cases the initial costs of the changes in terms of their annual equivalent plus the future running costs should be less than the existing running costs.

The reasons for alterations include:

- increasing the value of the asset
- more convenient or effective use of the building
- reduction in overall costs (maintenance, energy, wages, insurance, etc.)
- better appearance
- compliance with statutory requirements.

Significant savings can often be made by amending existing layouts within the structural limitations of a building. Studies of the movements of workpeople using such devices as string diagrams may show that a rearrangement of the work spaces and equipment would cut down the frequency and length of journeys between essential points. Assume that the total annual costs are broken down as follows:

	%
Buildings and equipment (amortised)	8
Maintenance, heating, lighting, cleaning, etc.	22
Staff salaries	70
Total	100

If movement accounts for one third of staff time the cost would be about 23% the total annual cost. A saving of 20% in the amount of movement would thus give about a 5% reduction in total costs – a saving which would be difficult to achieve in any other way.

For example, it has been estimated that changes in the layout of an industrial kitchen costing £10 000 would result in savings in labour and materials of £5000 per annum. It is required to determine the period of time over which the change must be satisfactory in order to justify the expenditure. The period for which the alteration should be satisfactory is given by the equation:

$$\frac{\text{Difference in running}}{\text{and operating costs}} = \frac{\text{Cost of alterations}}{\text{PV of £1 p.a. (or year's purchase)}}$$

In the example

5000 = 10 000 / PV of £1 p.a. factor

PV of £1 p.a. factor = 10 000 / 5000 = 2.0

It is then necessary to consult the standard tables for the PV of £1 p.a. and find the period that has a factor of 2.0 at the required rate of interest. Thus at 6% interest the factors are as follows:

for 2 years the factor is 1.833, and
for 3 years the factor is 2.673.

The period for a factor of 2.0 is between 2 and 3 years and is found by interpolation to be 2.2 years.
At 12% interest the factors are as follows:

for 2 years the factor is 1.690, and
for 3 years the factor is 2.402.

Again the period for a factor of 2.0 is between 2 and 3 years and is found by
interpolation to be 2.4 years.
Other methods of appraisal could be used for this sort of problem. For instance, the payback method would give a period of two years, which might be
sufficiently accurate for the purpose. Also, if the period over which the change
will be effective is known, the yield from the investment could be calculated for
comparison with other investment opportunities.

Conversions

The existing stock of buildings represents a considerable national asset and
clearly should be used as effectively as possible. The public discussion has usually centred around housing and the undesirable effects of breaking up established communities by the comprehensive redevelopment of slum areas. The
argument is that rehabilitation would not only be cheaper but would avoid some
of the social problems that beset new estates. However, the working environment is just as important and the same sorts of problems arise in the case of old
commercial and industrial buildings. In many cases by rearranging the internal
spaces and providing new facilities they can be given a new lease of life by making them suitable for some new use. This process is called by a wide variety of
names – adaptation, conversion, refurbishing, retrofitting, rehabilitation, renovation, modernisation, etc., representing the different approaches to the problem of making old buildings fit for new uses.

Basically the decision as to whether to adapt an old building or demolish it
and erect a new building will be based on the same sorts of economic analyses
as outlined earlier in this chapter unless the building is listed or in a conservation area or if there are other noneconomic reasons for its retention. Naturally,
consideration will have to be given to the physical condition of the building
and its probable life expectancy and whether or not the form of construction
will permit the necessary structural changes to be made. Also the size of the
spaces within the building and the ease with which they can be changed would
be important factors. Generally the old services will be inadequate and will have
to be completely renewed. However, refurbishing is usually a quicker process

than complete rebuilding and may offer other advantages, e.g. the old building might enjoy a higher plot ratio than would be permitted by the planning regulations for a new building on the same site or it might be thought desirable to preserve the character of the existing environment.

Energy management

The maintenance manager is likely to be involved in decisions concerning alternative ways of energy conservation. This will usually start with the preparation of an energy budget or audit giving the energy consumption of the building from all sources for internal environmental control stated in units of energy per unit of floor area. Comparisons can then be made of the energy usage of different areas and buildings to identify high cost areas. Further information is found in Chapter 11.

It is necessary, therefore, to carry out an investigation of the way in which energy is being used and to determine the measures that should be introduced to avoid waste. The measures include:

(1) Good housekeeping. This involves making the people who use the building aware of the need to conserve fuel by such simple measures as switching off the lights when they leave a room unoccupied. Perhaps it involves persuading cleaning staff that it is not really necessary to have all the lights on throughout the entire building for the whole of the period that they are working in the building at night.

(2) Reviewing standards. An examination should be made of the standards of heating, lighting, ventilation, humidity, etc., to make sure that they are not unnecessarily high.

(3) Modifications to existing systems. These may be quite minor modifications such as excluding draughts from windows and external doors or providing additional insulation. Others could be fitting thermostats to radiators or photocells to control lighting installations.

(4) Provision of new equipment. This would cover more ambitious schemes such as heat exchangers to recover heat from exhaust air or waste water, computerised controls to optimise the firing of boilers, and use of solar energy.

The financial consequences of the above measures can be determined using the appraisal methods described earlier. It should be mentioned that much more could be done at the initial design stage to provide energy efficient buildings by giving more thought to the basic shape, form and orientation of the building and by a more integrated approach to the design of the fabric and the services.

References

(1) Building Maintenance Information Ltd (1997) *The Economic Significance of Maintenance.* BMI, Kingston.

(2) Department of the Environment (1981) *English House Condition Survey 1981:* HMSO, London.

(3) National Institute of Economic and Social Research (1970) *Urban Development in Britain, 1964–2004.* Cambridge University Press.

(4) Department of the Environment, Transport and the Regions (1996) *English House Condition Survey 1996:* HMSO, London.

(5) Brand, S. (1994) *How Buildings Learn.* Viking Press, London.

(6) Stone, P.A. (1970) Economic Realities. *Official Architecture and Planning*, Feb. 131–4.

(7) Needleman, L. (1965) *The Economics of Housing.* Staples Press, London.

(8) Stone, P.A. (1980) *Building Design Evaluation – Cost-in-Use.* E. & F.N. Spon, London.

(9) Switzer, J.F.Q. (1963) *The life of buildings in an expanding economy.* Paper given at Royal Institution of Chartered Surveyors Conference, 1963.

(10) Chartered Institute of Building (1990) *Maintenance Management – a Guide to Good Practice,* 3rd edn. CIOB, Ascot.

(11) White, D.J. (1969) *Management science for maintenance.* Paper given at Conference on Building Maintenance, London 1969.

(12) Williams, B. (1994) *Facilities Economics.* BEB Press, Bromley.

(13) Bushell, R.J. (1979/80) *Preventing the problem – a new look at building planned preventive maintenance.* Institute of Building Information Service 11.

(14) Draft from a forthcoming Royal Institution of Chartered Surveyors Guidance Note: *Maintenance Procurement.* RICS, London.

(15) Flanagan, R., & Norman, O. (1983) *Life Cycle Costing for Construction.* Royal Institution of Chartered Surveyors, London.

Chapter 2
Maintenance Standards

2.1 Concept of standard

The identification of appropriate building standards is a key factor in determining the maintenance workload. The concept of an 'acceptable standard' is mentioned in a number of the definitions previously discussed, but this does not imply that there is any absolute standard that would be satisfactory in all cases. While some requirements may have universal applicability, e.g. structural stability, others would have to be assessed on economic and social grounds according to the overall policies of the owner or occupier. Robertson[1] names some of the factors which would have to be taken into account as 'political, legal and financial considerations, forecasts of future activity, market conditions, taxation policy and labour conditions'. Clearly the different requirements of different organisations will make different maintenance policies inescapable.

The dictionary definition of 'standard' as: 'object, quality or measure serving as a basis, example, or principle to which others conform or should conform …' indicates its role as a target set of conditions to which a given building must conform. Since there is no single measure of a building standard, this presupposes a number of subsidiary criteria that must each be met to achieve the overall standard. It is easy to talk of 'high standards' or 'low standards' in the context of some vague or undefined concept of quality, but the problem for the maintenance manager is how these standards may be objectively defined and performance measured against them. A further complication arises inasmuch as different people or organisations will differ in their perceptions of what constitutes an acceptable standard from their viewpoint; an issue which will be discussed in greater detail later.

It should be noted that the concept of a 'standard' relates to a function rather than an object. For example, a standard performance measure such as an adequate air temperature of 22°C in a building is independent of the means used to achieve the standard. In the foregoing example, various forms of heating (wet, warm air, gas, or electric) may be used. Also, the means of achieving this standard will be different in different circumstances; for example in tropical countries, attainment of temperature would involve refrigeration and air conditioning instead of, or as well as, space heating equipment, whereas in a cold northern climate it would not. Therefore the standard is independent of the means of achieving it, though the means themselves may be subject to a set of subsidiary standards. In the above example, the heating and/or air conditioning system would have its own standards relating to air changes per hour, boiler efficiency, thermostatic control etc, which themselves could be met by various different means. The whole concept of standards thus becomes a hierarchy, where the definition of an overall standard is achieved by specifying standards at the next lowest level that can achieve this. This is a key principle of the value engineering technique[3] in general, and of one of its most powerful analytical tools, the functional analysis system technique (FAST) illustrated in Fig. 2.1.

A further complexity of standards in maintenance is that they may change over time from the original provision. For example, bathrooms, a luxury a century ago, are now mandatory in new dwellings. Thermal insulation and standards of access for people with disabilities continue to rise over time. Other standards such as aesthetic taste and fashion change in less predictable ways. Therefore the overall environment in which building standards are set is fluid across different uses, users, scales, and time frames. A methodology for understanding this process in relation to continuing building performance is discussed next.

2.2 Standards and condition

It is attractive to think in terms of a range of acceptability, e.g. Kemp[2] suggests the fixing of upper and lower criteria as illustrated in Fig. 2.2. The higher limit

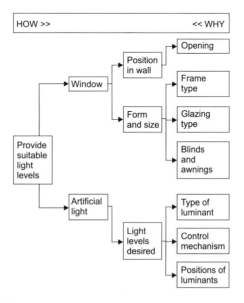

Fig. 2.1   The FAST diagram.

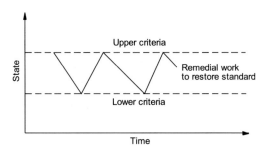

Fig. 2.2   The passage of time related to maintenance criteria.

is set by the cost of achieving it and the lower limit by the increasing probability of failure involving not only enhanced repair costs but also consequential losses where the normal user of the building is interfered with. The model is apparently related to functional performance and assumes a uniform rate of deterioration that will eventually result in failure. However, this is not an accurate representation of the behaviour of many building elements. Certain elements maintain a constant condition over the life of the building, while others are subject to sudden unpredictable failure. In some cases there is only a single criterion, e.g. whether or not the roof leaks, while in others there is no precise lower limit, e.g. decorations do not 'fail' in such a way as to impede user activities but must be judged on the basis of visual acceptance. Also, over the long life of a building, users will tend to come to demand higher standards and therefore repairs and renewals will inevitably contain some element of improvement. The model shown in Fig. 2.3 accords more closely with reality.

This generalised 'history' of a building element may be adapted to the characteristics of a particular element, as illustrated in Figs 2.4 to 2.9. A number of interesting variations will be noted.

In Fig. 2.4 the functional requirement of the paint film is static over time. Between repainting, the film undergoes periodic degradation. New technological developments in the later history produce a longer lasting, better finish coat than previously, giving an enhanced lifespan before the film deteriorates towards the lower level of acceptability; for example, peeling and blistering resulting in water penetration to the substrate.

The physical state of the installation degrades over time, but the functional requirements of the users change more rapidly; for example as the requirements for more sockets and appliances runs beyond the designed capacity of the original system. As the system drops to the lower functional limit, the resulting rewire will entail a large factor of improvement over the original design specification (Fig. 2.5).

In the variation in Fig. 2.6, two possible histories are illustrated. In the first (solid line), the roof deteriorates to the lower threshold before repairs and improvements are carried out after twenty years, to bring it up to the new standards (which may include thermal insulation when the original did not). In the second

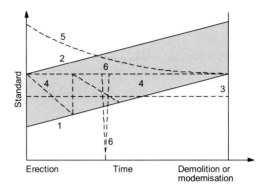

Fig. 2.3   Maintenance requirements related to standards over a period of time. (1) Lowest acceptable standard during time of use. The upward slope of the line symbolises change in the level of requirements. (2) Optimal standard during time of use; the area between lines 1 and 2 is the accepted standard area. (3) Element with immutable quality, i.e. horizontal line. (4) Rapid wear leading to maintenance during time of use, e.g. wallpaper. (5) Slow, undramatic change during time of use. Can be compensated by providing a higher standard from the beginning. (6) Dramatic failure calling for immediate action, e.g. leakage in water or sanitary installations. (Source: ECE (1969) *Proceedings of Seminar on Management, Maintenance and Modernisation of Housing*. United Nations, Warsaw.)

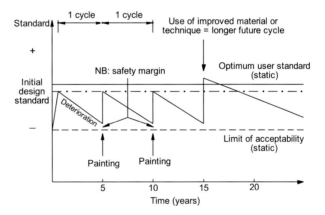

Fig. 2.4   Maintenance cycles – 5-year painting cycle.

history (dotted line), a localised failure (spike) results in an interim repair to the original standard, from which the covering deteriorates once more; though the repair is only satisfactory for 12 years or so before the roof covering's overall performance drops towards the rising minimum user threshold. The agencies of decay and changing standards remain the same in the two histories (hence the same gradients), but the intermediate repair intervention gives a different history for the repaired roof.

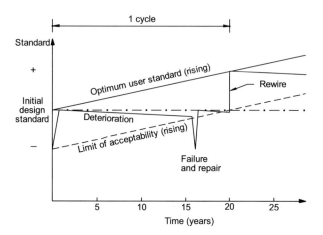

Fig. 2.5   Maintenance cycles – electrical system.

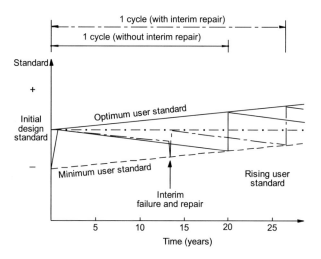

Fig. 2.6   Maintenance cycles – roof covering.

In the history illustrated in Fig. 2.7, there is only a small margin of acceptable performance – the system either works properly and fully or it is deemed unsatisfactory from a health and safety point of view. This points towards frequent, regular minor interventions in the form of yearly services. A passenger lift installation would show a similar pattern over a six-month cycle.

In contrast to the preceding history, in Fig. 2.8 there is a large margin of acceptability for the state of the brickwork pointing before it has a marked effect on the weather-tightness and structural stability of the wall itself, though it may look worn and unkempt long before that point is reached. The acceptability level remains static over the long lifespan. This is an example of a repair which because of slow deterioration rarely becomes critical, as it is often 'bundled' with

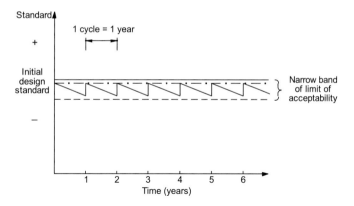

Fig. 2.7   Maintenance cycles – fire defence system.

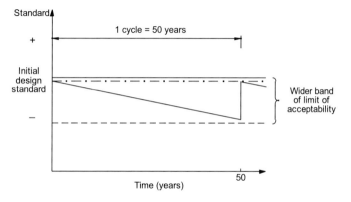

Fig. 2.8   Maintenance cycles – brickwork repointing.

other repairs in the area before the critical state is reached, or else the building itself reaches the end of its useful life and is refurbished or demolished.

In contrast, communications systems show such a rapid technological advance that the equipment becomes obsolete long before the end of its physical life is reached. Two histories are shown in Fig. 2.9, one system (dotted line) has a longer physical life than the solid, but this results in very little extra functional life because of the rapid pace of obsolescence. Also shown in the dotted-line system's history is the effect of obsolescence inasmuch as for a period before replacement, it may no longer meet functional requirements. This illustrates that for certain building components, long-life is wasteful of resources, and a cheaper short-life component would be more cost efficient. No one buys a telephone or a pot of paint designed to last fifty years.

Having considered the particular requirements for each individual element, the standard may be expressed in the following ways.

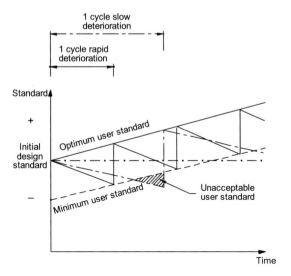

Fig. 2.9   Maintenance cycles – communications systems.

Physical terms relating to:
(1) The condition of the element specifying the magnitude of the defects which calls for remedial action. Wherever possible, the extent of the defect should be capable of direct measurement to ensure uniformity of interpretation when more than one person is responsible for the inspection.
(2) The performance of elements or environmental systems. Where performance specifications have been used for the initial design, these could be adopted for this purpose.

This system is called 'condition controlled maintenance' and it presupposes that there will be inspections at appropriate intervals in order to determine by visual means or measure whether or not the condition of the elements or their performance has deteriorated below that laid down.

Times at which repairs and replacements are to be made
This method is sometimes referred to as 'frequency based maintenance' and it requires certain knowledge of the rate of deterioration and of the point in time when either functional failure is imminent or the appearance will become unacceptable. Clearly a proper balance must be achieved between the frequency and the risk and consequences of failure. An illustration is given in Figs 2.10 and 2.11 in relation to lamps in a large building such as an airport or a hospital. Figure 2.10 shows a histogram of a population of bulbs in such a building, with a mean failure time of 2000 use-hours. It can be seen that the failure rate is initially low, but as the average lifetime approaches, failures become more frequent and rise to a peak, before falling off as the remaining bulbs' actual lifespan is reached. The second graph is a cumulative measure of the same data, which

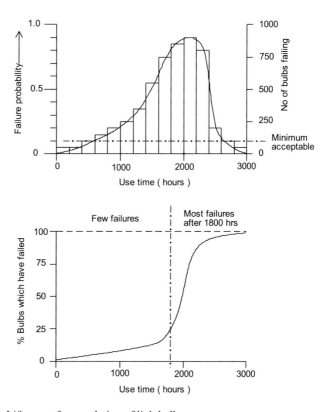

Fig. 2.10   Lifespan of a population of lightbulbs.

shows failure frequency rising from a low level initially, over a comparatively short period.

If replacements of these bulbs were carried out on an *ad hoc* basis as a reaction to failure, the maintenance profile would be as shown in the uppermost graph of Fig. 2.11. For the first few failures, replacement would not be necessary as the remaining bulbs would continue to provide the required standard of illumination; which standard would be a little below the design standard in order to allow for such a margin of failure. The first replacement, P1, is therefore a 'catch-up' programme. Thereafter, as the bulbs fail increasingly frequently, more frequent and larger interventions are required to maintain the required lighting standard. This subsequent pattern, P2 to P8, approximates to the bell curve of the failure rate preceding but becomes progressively blurred as early failures of replacement bulbs start coinciding with late failures of the original bulb population (P7, P8). Extended, the *ad hoc* regime results in a more or less random failure pattern and a correspondingly even and constant demand for replacements in order to maintain the desired level of luminance.

In the second profile, the replacement pattern anticipates the failure pattern in order to limit interventions. Instead of a number of *ad hoc* replacements, the first partial replacement is a catch-up on early failures, necessary to retain the

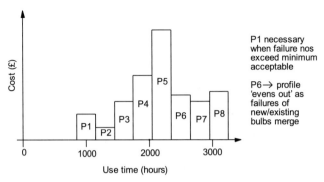

P1 necessary when failure nos exceed minimum acceptable

P6→ profile 'evens out' as failures of new/existing bulbs merge

Proportion of 'P' cost = Transport/setting up and user distribution

Maintenance profile as a reaction to failure

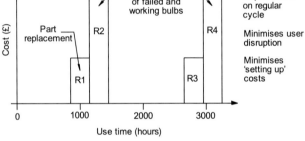

Replacement on regular cycle

Minimises user disruption

Minimises 'setting up' costs

Maintenance profile in anticipation of failure – preventative

Fig. 2.11 Maintenance profile.

lighting standard (R1). In anticipation of the failure most of the rest within a comparatively short time, a wholesale preventive relamping is carried out (R2), the result of which is to cause the entire population of bulbs to revert to the 'as new' situation. Although this regime is seemingly wasteful inasmuch as some perfectly functional lamps are being discarded ahead of failure, in terms of minimising disruptions to the users, preserving the minimum standard of luminance, and achieving economies of scale in terms of labour deployment and bulk acquisition of replacement lamps, the second, planned, anticipatory approach is in fact the more efficient.

Financial criteria

These may take the form of a variable sum related to the cost of some primary activity or replacement value, or a fixed sum based on historic costs, or an analysis of anticipated benefits.

In most cases, all three methods are used, although there is little doubt that financial expediency often takes precedence over the physical needs of the building. A possible reason for this was revealed in a study carried out by Bath University in which it was found that in the larger firms the initial technical assessment was carried out at a low level in the firm, but that the ultimate decision as to how much to spend on maintenance was taken at a higher level and was in many cases little more than a trimming exercise. The budget was generally the amount expended the previous year with, perhaps, some adjustments for increased costs. While not denying the need for financial constraints, there should be greater awareness of the ultimate cost of under-maintenance. Thus instead of thinking in terms of the current availability of funds the total costs over a substantial period, say ten years, of alternative maintenance policies should be compared.

The life cycle method of calculating and comparing these ongoing costs was discussed in Chapter 1. On the scale of a building or estate as a whole, the profile for maintenance and running costs compared to the initial capital cost of the building may be represented as follows in Fig. 2.12.

It will be seen that as the building ages, the two factors of cumulative physical deterioration and increasing obsolescence tend to force up the maintenance budget on a year-on-year basis. Cleaning and running costs tend to increase at a more or less constant rate, whereas the maintenance expenditure is concentrated more in 'spikes' which relate to particular replacements or refurbishments, carried out to remedy elements at the end of their physical life and those which have become obsolete. These 'spikes' of expenditure tend to become larger and more frequent over time, assuming a constant building use (i.e. it does not change ownership and use to one requiring a lower functional standard). The discounting techniques of present value (PV) for one-off costs such as replacing a roof covering, and year's purchase (YP) for ongoing costs such as cleaning, enable these periodic costs to be assessed and compared as costs at a single point in time; for example related to the capital cost of the building, as shown in Fig. 2.13.

From this type of analysis, various running and replacement strategies may be evaluated against a common base to determine the most efficient route to

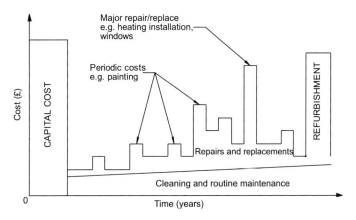

Fig. 2.12 Capital and replacement costs.

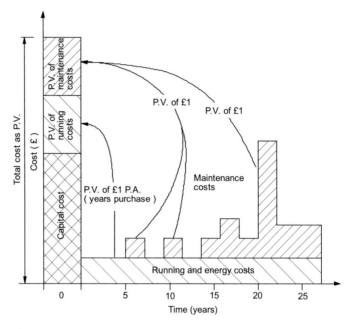

Fig. 2.13   Total cost analysis.

maintaining the required standards. For example, the above analysis may be used to evaluate whether the extra capital cost of uPVC windows over cheaper softwood timber windows would be justified in terms of the savings on periodic painting costs over the lifetime of the windows.

2.3 Standards generally

Having examined the relationship between acceptable standards and perform-ance, it is necessary to look in more detail at how the standards of maintenance are agreed. Generally, the achievable standard will represent a balance between need and resource, with the provider or controller of the resource usually hav-ing the greatest influence over what this balance may be. However, the owner's/provider's view of what may be an acceptable standard often differs from that of the users, or indeed from other interested parties such as statutory housing or health and safety inspectors, or even the general public if they use the building. As Chanter and Swallow[4] put the point: 'It is important to distinguish between the ability to pay and the willingness to do so. To create maintenance activity both must exist'. This tension between provision and use runs throughout build-ing maintenance operations. For example, a surveyor may carry out a condition survey, and arrive at a schedule of repairs that he or she considers necessary based on the inspection of the physical state of the building. However, there is frequently a divergence between:

- the amount that the surveyor would like to spend on the property to put it into first-class structural and decorative repair, and
- the amount which it is considered economic to spend on the property in view of its profitability.

If there is gross disparity between the amount based on the physical condition of the building and that based on financial considerations, the future of the property is considered. The main disadvantage would seem to be that the least profitable premises are likely to be the most dilapidated and yet progressively less and less money will be spent on their repair. Also, the surveyor bases the extent of the work on a subjective notion as to what constitutes a first-class condition. In many cases economic considerations will dictate a lower standard and therefore conflict between the two amounts is inevitable.

2.4 Standards from different viewpoints

In practice much will depend upon the attitude and status of the person responsible for initiating maintenance work. The primary initiators of maintenance are the owner and/or occupier, although other interested parties, e.g. health and safety inspectors, insurance companies, employees and their trade unions, members of the public, etc., may exert either a direct or an indirect influence on the amount of work undertaken.

Owner

The attitudes associated with the owner function will be broadly similar whether the owner is in occupation or leasing the building. The primary aim is to preserve the value of the asset so as to ensure a long-term trouble-free investment capable of providing a continuous and satisfactory return. The object is to achieve this with the minimum expenditure. Ideally, advance provision should be made for future repairs that should be preplanned so as to achieve an even cash flow over the years.

However, if the building is viewed purely as an investment then clearly the return should be no less than that obtained from alternative investment opportunities. In the absence of current or future benefits, there is no economic incentive for the owner to do more than is necessary to comply with statutory and contractual obligations. Also, the owner's long-term interest in the property leads the owner to place greater emphasis on structural repairs rather than the internal decorative condition which has little direct effect on the rate of deterioration of the property. However, there may be indirect consequences in that the degraded internal appearance may result in less careful treatment by the occupier.

Occupier

The attitudes of the occupier are distinguished here from those of the owner although both would be present in the case of an owner/occupier. Clearly the maintenance policy must be related and subservient to the fundamental aims and objectives of the individual or organisation which occupies the building. In order to achieve its objectives an organisation must initiate and sustain certain activities or modes of behaviour that require the provision and maintenance of a compatible spatial and physical environment. The maintenance needs of the organisation may therefore be stated at four levels of specificity:

(1) The effect of maintenance standards on the objectives of the organisation. The main aim of most organisations is survival, which in a commercial context is dependent upon making a profit. However, there is a growing need for criteria which go beyond simple economic analysis and take into account the effectiveness of maintenance in relation to corporate objectives broader than that of profit maximisation or which reflect more sophisticated ways of ultimately achieving maximum profit. The standard must therefore reflect the social attitudes towards the environment in which people live and work and is only partly determinable on a rate-of-return basis.

(2) The effect on the activities necessary to achieve the desired economic or social objectives. This is relatively straightforward where the purpose of the building is to house manufacturing processes which require rigorously controlled conditions of temperature, humidity, dust particles, etc., in that deviations from the 'norm' may have measurable effects on output. Human beings are much more adaptable and their reactions to changes in the environment not entirely predictable except in extreme conditions. An understanding of the physiological and psychological factors underlying human sensation and perception is necessary in order to predict the effects of marginal changes in environmental conditions on user activities.

(3) The effect on the internal environment in which the user activities take place. This involves consideration of the total environment produced by the building, the people and process machinery that it houses and the external environment. The cost of maintaining the environmental conditions at the required level may be significantly affected by deficiencies in the state of the building fabric. For example, the greater heat loss through damp insulation caused by a roof leak must be made good by the input of heat with correspondingly higher fuel costs.

(4) The effect on the physical condition of the building elements and materials of construction. In particular, the long-term effects of defects on the element affected and on adjoining elements should be considered.

At present the initial technical assessment is carried out at level 4 and expenditure limits at level 1, but little attempt is made to trace the effects of disrepair through the intervening levels. The result is the apparent conflict between the maintenance organisation and upper management in terms of resource allocation. This is dealt with further in Chapter 5.

Specific performance standards for analysis at level 3 exist in codes of practice, regulations and research recommendations, although in many cases these appear to be based more on preconception and intuition than on a scientific study of level 2 criteria. The weak link in the chain is undoubtedly lack of knowledge of the way in which the environment affects the behaviour of people. Without this knowledge, it is difficult to relate maintenance standards precisely to the degree of fulfilment of objectives.

Division of responsibility

A property owner normally seeks to preserve the condition of his or her property by the insertion of appropriate clauses in the lease or rent agreement. However, in some cases the position is regulated by statute. Thus, Section 8 of the Landlord and Tenant Act 1985, provides that in the case of certain small dwellings let for a period less than three years there shall be an implied term in any contract of letting that the house is at the commencement of the tenancy 'fit for habitation' and shall remain so during the tenancy. The criteria for determining fitness for habitation are as stated in Section 4 of the Housing Act 1957 (now re-enacted in Section 604 of the Housing Act 1985).

Section 11 of the Landlord and Tenant Act 1985, provides that in any lease for less than seven years the property owner shall:

(1) keep in repair the structure and exterior of the dwelling (including drains, gutters and external pipes),
(2) keep in repair and proper working order installations for the supply of water, gas and electricity and for sanitation (including basins, sinks, baths and sanitary conveniences but not other appliances that make use of the supply of water, gas and electricity),
(3) keep in repair and proper working order the installations for space heating and heating water.

The standard required is a variable one based on the age, character and prospective life of the house and the locality in which it is situated. The condition of the other houses in the area of a similar type is probably the main criterion used when determining an appropriate standard.

At the moment, there is no similar code imposing specific responsibilities for repair and maintenance upon property owners and tenants of other types of property. The usual basic terms in a modern commercial lease require the tenant to keep the building in good and substantial repair and to redecorate both the interior and exterior at certain intervals, usually three to five years for the exterior and five to seven years for the interior. It should be noted that the expression 'keep in repair' also means 'put in repair' and that it is a wise precaution to see that a schedule of condition is attached to the lease in order to limit the liability both during and at the end of the lease. Generally such a repairing covenant does not impose a liability to provide something which was not there before. Thus, if there is no damp-proof course, lessees are not obliged to provide one although they would be responsible for remedying defects that arise because of this omission. A more onerous provision is that the lessee should be liable for 'inherent defects' which would extend to making good design defects. Formerly the expression 'fair wear and tear excepted' was commonly used in leases for short terms. This has been legally interpreted as defects caused by the normal action of the elements or the normal human use of the premises. However, the lessee is bound to do such repairs as may be required to prevent the consequences flowing originally from wear and tear from producing others which wear and tear would not directly produce.

Where buildings are let in multiple occupation, it is normal to require tenants to pay a service charge to cover among other things the maintenance of the structure, the exterior and common parts of the interior for which the property owner is responsible. Such a charge should be subject to review at fairly frequent intervals so that adjustments can be made for increases in the cost of carrying out the necessary work. In the case of flats the Landlord and Tenant Act 1985, provides that the costs must be reasonably incurred and that the services or work must be of a reasonable standard. The Act further protects the interests of tenants by giving them a right to require the property owner to supply a summary of the costs incurred and to inspect the accounts and receipts on which the summary is based. Where tenants have a liability as part of the service charge to pay for repairs they have the additional right to be consulted before major works are carried out by the property owner, who is required to obtain at least two estimates for the work.

An important factor is that the lease provides a means whereby the property owner can impose a system of planned maintenance upon the tenant to ensure that the condition of the property does not deteriorate. The provisions can be enforced by reserving a right of entry to inspect the property during the period of the lease and by requiring the premises to be delivered up at the end of the lease in the same condition that they were in at the beginning of the tenancy. In order to enforce the requirements of the lease the property owner may serve on the tenant a 'schedule of dilapidations'. There are two types of schedules:

(1) *Interim Schedule of Dilapidations* served during the currency of a lease together with a notice to repair. This is a list of all the items of repair necessary under the terms of the lease and is usually served with the intention of enforcing a right of re-entry if the tenant fails to carry out the work. Some relief is provided in the case of residential property under Part 2 of the Landlord and Tenant Act 1954, where there are more than three years of

the lease remaining unexpired. In this case tenants may apply to the court for relief and property owners must show that in the absence of compliance they would suffer an immediate damage to their reversion. There must be a specific clause in the lease document permitting the serving of such interim schedules, such as the requirement to 'keep in repair' rather than the more usual 'leave in repair' clause, which triggers the end of lease schedule.

(2) *An End of Lease Schedule of Dilapidations* which forms the basis of a money claim against the tenant in respect of breaches of covenant to keep and leave in repair. This is similar to the interim schedule but the items of repair are priced out to form the basis of a lump sum claim for compensation in lieu of the tenant carrying out the repairs. Surveyors should note that it is merely a costed list of work required under the terms of the lease, and is not a specification of the manner in which the repairs should be carried out. The costs are payable unless they would exceed the benefit to the lessor on reversion; for example on a property that has been vacated at the end of a lease and which the lessor is planning to demolish as part of a redevelopment.

It is important for the maintenance manager to be fully aware of any divisions of responsibility and the attitudes of the parties when determining appropriate maintenance standards. The minimum standards will always be those required under statute (see Chapter 3), which generally relate to basic health and safety provisions ranging from fire defence to effective sanitation. Above these basic requirements, the appropriate standard is that which is adequate for the provider in view of resources available. However, the maintenance manager has a duty to inform the provider of the consequences of deferred maintenance in terms of cost growth and deteriorating environments, but this should not take the form of strident or alarmist statements of dire consequences which are unlikely to be fulfilled in the short term, but rather as a measured assessment of the realistic long-term consequences.

2.5 Standards and performance

The factors which influence decisions to incur maintenance expenditure are complex and in some cases conflicting. However, the factors should be made explicit in a planned maintenance policy spanning a number of years. The factors include the satisfaction of user requirements, value considerations and statutory constraints. Each organisation will thus have its own standards based on its particular needs and level of resource available. However, the criteria for these standards have a common basis, a framework for which is given:

Maintenance and occupation standards: criteria

The following high-level criteria relate to broad functional requirements of buildings in use. Their applicability in the context of maintenance is that these provisions are subject to loss and deterioration in use and therefore require maintenance operations to keep them at the required operational level throughout the building's life.

Quality of internal space

(1) *Adequate amount of space available to user.* Although this is primarily a design criterion, the occupation pattern of the building will determine whether there is sufficient space for the various uses and occupants. As the occupation pattern changes over time, and may have departed radically from that envisaged at the design stage, it comes within the sphere of maintenance management in ensuring that the building continues to be suitable for its function. In the case of overcrowding, it is essential that the maintenance manager remains aware of how this may compromise the ongoing provision of essential services such as means of escape in case of fire.

(2) *Structural stability.* The freedom from collapse is perhaps one of the most constant and unambiguous standards across estates and users. An example of how this is further subdivided into more specific performance requirements is given below.

(3) *Shield from the external environment.* The degree to which the external envelope shields the users from the extremes of climate and other outside influences such as traffic noise and pollution. This is not an absolute requirement by any means: some degree of controllable interaction with the outside environment is necessary or even desirable. The most obvious example is that of opening doors and windows, where too efficient a shield, such as in an air-conditioned building with no openable windows, can lead to internal environmental problems such as sick building syndrome (see Chapter 11). This will also include freedom from dampness and other potential environmental health hazards.

(4) *Sufficient natural light and air.* With the notable exception of a few specific uses such as cinemas or operating theatres, most buildings require the maintenance of adequate natural lighting by means of windows, lights, etc., which have a consequent maintenance and cleaning component. Note also that at some times of the year, and in other countries, such as the Middle East, it may be desirable to limit or control the amount of light entering the building.

(5) *Adequate access, signage, security, and circulation.* The maintenance of suitable access controls into and within the building is important for both general utility, and health and safety, for example, adequate means of escape. Changes in standards, for example recent changes in requirements

for access for people with disabilities, may require alterations of existing, previously satisfactory provisions.

Quality of services

(1) *Adequate sanitary and washing facilities.* The maintenance of working sanitary and washing facilities is frequently one of the highest priorities for the maintenance manager, since malfunction or loss of service has immediate and far reaching consequences on the use of the buildings and the health of the occupants.

(2) *Adequate space heating/cooling.* Similarly, an important requirement for the ongoing use of the building.
   - adequate cooking and food preparation facilities
   - sufficient rest and recreation facilities
   - adequate communications

The degree to which the final three categories above are applicable will depend largely on the use pattern of the building. Certainly, computing and telecommunication facilities are increasingly perceived as essential services whereas in the past they were not.

General

(1) *Health and safety.* Although implied in many of the above categories, modern practice requires the systematic assessment and minimisation of all potential risks within a building. This function is frequently carried out as part of the maintenance management operation.

(2) *Aesthetic, fashion, presentation, prestige.* Most decorations are replaced not because of physical deterioration but because of the desire to change the existing provision. The type and quality of decorations and finishes send out important messages to the users of the building, and are the building's most visible manifestation to the users. Frequently, laypersons will judge the quality of a building (and the quality of the maintenance management operation responsible for it) solely on its appearance, i.e., its decorations and finishes, rather than on its underlying state and suitability. Replacement of these, including the mechanisms for the choice of new colour, style, etc, is an important component of maintenance management.

(3) *Comfort and luxury.* Independent or additional to the functional efficiency of the building, the 'feel' and perceived quality or exclusiveness of the finishes, fittings, and services are an important attribute in many buildings. For example, a Michelin star restaurant will have much more luxurious chairs, tables, floor coverings, lights, wash hand basins, etc., than a more modest cafe, even when superficially their function is similar.

(4) *User-defined ...* Other standards may relate to the specific circumstances of a particular user.

These high-level standards may be factored down into more specific standards relating to each, using the FAST methodology outlined earlier. For example, the high-level standard of 'structural stability' may be factored as follows:

- structural stability:
  o adequacy of foundations
  o freedom from serious cracking in masonry, concrete, etc
  o freedom from bowing and distortion of structural walls
  o safe loading and function of joists, beams and lintels
  o adequacy of lateral restraint
  o adequacy of walls and roof against wind and snow loads
  o adequacy of structure to resist applied loads
  o adequacy of structural performance in case of fire

etc.

These factors can be subdivided further to form the basis for the condition inspection of a building, discussed in more detail later in this chapter.

2.6 Priorities and criticality

For particular estates and uses, the standards may be prioritised in order of importance to the running of the occupying organisation. For example, Bushell[5] gives the broad priorities for healthcare buildings as:

(1) safety
(2) essential service
(3) statutory requirements
(4) security
(5) initial cost
(6) revenue saving
(7) spares availability
(8) alternative source of supply
(9) delivery time
(10) human resources
(11) public relations

Whilst such lists are useful in determining overall policies and priorities, in the actual inspection process which precedes maintenance work, a more universal measure of acceptability is appropriate. Stevens[6] draws attention to the difficulty of making rational decisions about the appropriate level of maintenance expenditure. He proposes that assessments of building condition should be made on a numerical basis and divided into five classes ranging from very good to dangerous. Similarly, he suggests that the desirable level of maintenance should be expressed in five classes ranging from very high to very low according to the use of the particular area of the building. A comparison of the actual condition of the part of the building under consideration and the desirable level of maintenance for that part will indicate the priority to be accorded to the work.

Limiting this to the habitable or working areas of a building it is possible to distinguish three broad levels of intent:

- *Lowest level.* The immediate environment should not be directly harmful or uncomfortable.
- *Middle level.* The environment should be such as to minimise effort at work or maximise output.
- *Highest level.* The environment should promote not only the actual wellbeing of the users of the building, but also their sensation of wellbeing.

Lowest level

Achieving the lowest level was the task of the nineteenth century reformers who were in the main medical men rather than architects. The bringing together of large numbers of people and the pollution of the atmosphere by industry gave rise to environmental problems of the most compelling urgency. As early as the 1860s the difference in health of those working in controlled environments, even if crudely controlled, and those working in relatively uncontrolled ones was a matter of public record. The public health concepts developed during this period form the basis of much of our presentday legislative controls.

Middle level

The early reformers were thinking purely in terms of health and safety, but clearly if conditions were so inadequate that they increased the sickness and accident rate, productivity would also suffer. Most studies have centred on specific occupational hazards or on extreme conditions and have shown correlation between temperature and time lost through sickness and between temperature and the accident rate. All the studies show that workers are much more sensitive to nonideal conditions of temperature and air movement than to other environmental factors for which the range of acceptability is very much wider. Experiments devised to give precise numerical expression to standards suffer from the fact that the environment, even under test conditions, is made up of a complex interaction of factors and it is quite impossible to expose the subject to only one factor at a time. However, some guidance is given by studies in which the subject is required to carry out a standardised task and the performance is assessed under varying environmental conditions of temperature, noise, lighting, etc.

Highest level

At this level psychological factors play an important part, i.e., the interplay between the subject's immediate intake of information from the surroundings and that information which is stored in his or her memory from previous experience. A host of factors determine the probability that a particular stimulus will be transmitted through a sensory channel – the nature of the stimulus, existing events taking place in the sensory system, the arousal level of the individual, previous experience of the stimulus, etc. The incoming signal is transformed from a sound wave, light quanta or heat into a series of neural pulses that travel to the central nervous system. There is an interplay between the incoming data and the information stored in the memory after which a response takes place. The response may be an active one, e.g. distraction by noise, dazzle by light, shivering in a cool room, or a passive response in which the *status quo* is preserved.

Studies in this field show the difficulty of establishing universally acceptable standards at this level of sophistication. Not only will people select different stimuli, but they will interpret them differently according to their past experience. One would expect therefore a fairly wide range of preferences according to the age, sex, ethnic background, habits, etc., of the person involved. Thus, rigid controls based on narrow and probably short-lived concepts of what is desirable are inappropriate – the aim should be to provide as wide a range of options as possible. This would also suggest giving individual users a greater voice in the standard of their immediate environment rather than laying down a uniform standard.

More recent studies[7] have indicated that in terms of condition assessments, the 5-point scale or some close variation is very common practice in building maintenance. A typical scale, taken from the Appendix, is as follows:

Physical condition

This criterion assesses the physical state of repair of the element.

C0: not inspected/recorded; assumed satisfactory
C1: as-new condition
C2: fully functional, showing some signs of wear
C3: function slightly impaired in some ways, but still operational
C4: function deteriorating and requiring attention soon
C5: nonfunctional or absent
C6: dangerous

The C1 'as new' category is the starting point for a new element in a new building, from which point subsequent deterioration is measured.

The C3 and C4 assessments may be suffixed by / and a number indicating the time in years to anticipated failure, e.g. **C4/1** = within 1 year.

'R' may be suffixed to indicate replacement as only option (i.e. the element cannot be repaired *in situ*), e.g. **C4R/1**.

The C6 category is intended as a specific highlight for very urgent or emergency work, on which immediate action is required.

The C0–C6 levels are sufficiently specific for recording most maintenance needs. However individual users may specify the codes in more detail, using a decimal point, if a finer 'grain' is required.

A supplementary scale, recording obsolescence or deviation from design standard, may also be used:

Obsolescence

This criterion assesses user need versus current provision

N0: Not inspected/recorded; assumed satisfactory

N1: Element satisfactory for present use

N2: Element functional, but falls short of new product/installation

N3: Element is not adequate for user needs by current standards

N4: Element is wholly inadequate for users and seriously affects their use of the premises.

The use of these scales begs the question of what is assumed as 'satisfactory' for a particular use. This is a very important question, as different assessors, even from within the same organisation, have been shown to have very different perceptions when identifying deviation of building components' performance from standards[8]. This highlights the importance of adequate briefing of surveyors inspecting buildings, to ensure that they are aware of the definitions of 'adequate' as they apply to the particular estate they are surveying. This is frequently ignored in practice, with the result that either actionable repairs are given insufficient priority; or more commonly, in order to play safe and cover potential professional liability, that unnecessary work is prioritised for action, giving rise to the perception of 'over-maintained' buildings referred to by Williams and quoted in Chapter 1. At the point of survey this tendency to overestimate necessary repairs is almost impossible to detect, and indeed may only become manifest, if at all, in inflated repair costs on consequent planned maintenance programmes.

Criticality

Condition ratings similar to those above are often applied to the whole of the building, thereby by implication applying the same standards throughout. Yet it is common knowledge that some areas of a building are much less important in terms of one or more of the applicable standards, than other parts. For example a sales area of a store would need to be maintained to a much higher standard than the outside shed used for low-grade storage. Or the boardroom would require a higher standard of finish than the tearoom. Yet, to simplify data collection, this fundamental difference in applicable standards is frequently overlooked in condition assessments and prioritising repairs, with the consequent misdirection and waste of a resource far greater than that saving 'achieved' by the simplification in data collection procedures. To take these different standards into account, some organisations have adopted various measures of criticality assessment, whereby each area is given a simple code according to its comparative importance, or criticality, to the functions of the occupying organisation. A further related measure is an assessment of the impact of failure of a particular component on the use of the building. A method of criticality assessment particularly suited to automated data processing is given as follows, taken from the Appendix: Standard Maintenance Descriptions:

The Index of Criticality is a classification of the levels of maintenance appropriate to particular areas or buildings according to their relative importance to the building user.

There are two components:

- the *location* component
- the *use* component

Both are rated from 1 to 5

Location index

Level **L1** – (very high) operating theatres, computer areas
Level **L2** – (high)   boardrooms, reception, prestige areas
Level **L3** – (standard)   main office areas
Level **L4** – (low)   sheds, outbuildings, low-grade ancillary
Level **L5** – (very low)   prior to demolition; short-life

Use index

Level **U1** – (very high) health and safety related, fire defence, etc.
Level **U2** – (high)   failure would halt user operations
Level **U3** – (standard)   failure would be costly, or inhibit users
Level **U4** – (low)   failure would be moderately costly, etc.
Level **U5** – (very low)   failure would be a minor nuisance

The *Combined Criticality Index* (CCI) based on the importance and status of the element or area within the estate, plus the consequences of element failure and/or cost growth of repair postponement, is arrived at by multiplying the Location and Use Indices, giving a range of 1–25. For example:

The **CCI** of a broken window in a main office area could be: **L3** (standard location) × **U1** (health & safety hazard) = **3**.

The **CCI** of a leaking gutter in a garage could be: **L4** (low-grade location) × **U4** (moderate nuisance) = **16**.

The weightings of the location and use indices, here given as 1:1, may be varied according to the needs and priorities of the particular user.

Action thresholds and priorities may then be assigned according to the CCI. The lower action threshold CCI may be set at 15, i.e., anything greater gets actioned as and when resources are available.

This CCI, coupled with the observed state of repair of the element, provides the basis for prioritising repair requirements for a property estate, and could be determined according to the particular requirements of the maintenance organisation at that time without compromising the standardisation or quality of the condition data collected. One significant advantage of this method is in the actioning of prioritised repair lists inasmuch as the consequences of increasing or reducing the allocated resource may be immediately seen in detail (i.e. which specific repairs have been actioned and which postponed). Also, indices for criticality can be changed or remodelled later, without invalidating the accuracy of the survey data in any way. This permits the modelling of complex 'what if' scenarios and thus enables a more sophisticated maintenance need assessment.

2.7 Condition monitoring

The most common method of measuring the compliance of buildings with their required performance standards is during condition assessments. These assessments can be undertaken at a range of levels from the 'broad brush' to the specific.

Seven types of condition surveys are defined:

(1) cursory inspections, broad-brush appraisals, and evaluations
(2) structural condition surveys
(3) cyclical surveys as part of a planned maintenance programme
(4) specific detailed surveys to prioritise and schedule repairs
(5) surveys to compare building and services provision and condition to user needs
(6) specialist surveys such as dampness investigations
(7) fault check or specific failure surveys

although a particular survey may combine several of these types.

The survey type should be chosen and structured to provide sufficient and sufficiently detailed information to fulfil the specific goals of the survey.

Extraneous and irrelevant information should not be collected, as this will divert resources and obscure the relevant data.

The purpose of the surveys may range from the very broad; for example to assess an estate of buildings on acquisition to prioritise further action, to the specific logging of wants of repair prior to actioning works orders. Increasingly, the process is automated by the use of handheld computers, and it is probable that within a few years this will be overwhelmingly the predominant method of data collection. However, the pitfall in using such automated systems lies in the assumptions that may lie behind the programming of the system. Often, the original structure and program was written some time previously; perhaps (in the case of 'off the shelf' systems) for another organisation; yet the assumptions about relative standards in respect of what to survey, in what detail, and to which assessment regime which lie behind the original programming may be applied to a particular user whether or not they are relevant. For example, a survey format may have insufficiently detailed data fields to collect required information on air conditioning, having originally been designed for housing surveys. Conversely, the survey format may request data irrelevant to the particular user and/or purpose of the survey. Consequently, the maintenance manager must carefully evaluate these assumptions and characteristics of an automated condition analysis package before deploying it, in order to avoid collecting distorted or irrelevant data. Ideally, customised packages that enable user-defined menus for assessment are to be preferred.

As a final point about condition surveys, those undertaking the survey must be given training not only on how to enter and process the data, but also on the criteria and standards of the organisation which underpin the survey, to ensure accuracy in applying the organisation's standards in evaluating the building's actual performance.

2.8 Value considerations

The value of a building is determined by the demand for the services that it offers in combination with other factors of production. Thus in the absence of demand a building has no value to be maintained and neither the initial cost nor the standard of maintenance has any economic significance. Of course, if the lack of demand is purely temporary and it can be foreseen that there will be a future demand for the building then maintenance expenditure can be related to the anticipated future returns discounted to present values.

Assuming, however, that a demand exists, the question arises as to how far the condition of the building would affect the price that would be paid for its services in the open market. This can only be assessed by determining the relationship between building condition and user activities. The difficulty lies in separating the contribution made by maintenance from that of the many other factors. Maintenance interacts with other costs and also with revenue, e.g. for shops it is assumed that a higher level of maintenance will draw more customers and induce them to buy more or pay higher prices, while in manufacturing the standard of maintenance may affect the output of the workers and the quality of the product.

It has been proposed that the optimum level of expenditure on maintenance is that which gives the maximum return (see Fig. 2.14). The assumption is that there is an incremental relationship between maintenance expenditure and the value derived from that expenditure and that each additional increment in maintenance expenditure produces a progressively smaller increase in value.

While this would appear a convenient analytical model, it does not accord with reality in that it presupposes a precise and calculable relationship between the standard of maintenance and the efficiency of user activities. It is a matter of common observation that many activities show a high degree of tolerance to the surrounding environment and that quite substantial changes in the condition of the building have little or no effect on them. It is necessary, therefore, to identify:

(1)   Those user activities which are sensitive to the physical condition of the building and to isolate some aspects of those activities which display measurable changes under different environmental conditions, e.g. speed or accuracy of working.
(2)   Those building elements which play a significant role in providing the necessary conditions and to assess their probable rate of deterioration and the cost growth of repairs.

Thus, where there is a relationship between building condition and user activities, the consequences of delaying maintenance can be made explicit. In some cases, the loss in value (decrease in user efficiency) may be immediate and progressive, as shown in Fig. 2.15a, while in other cases the loss may not occur until deterioration is far advanced as in Fig. 2.15b. Where the benefits are long term the expenditure incurred on preventive maintenance is not very different from

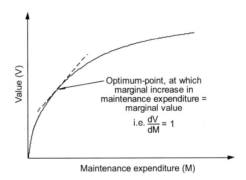

Fig. 2.14   Maintenance expenditure related to building value.

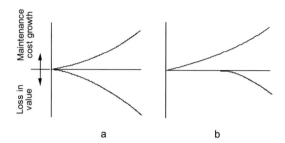

Fig. 2.15   The effect on building value of delayed maintenance.

that incurred in the initial erection and is more in the nature of capital than current expenditure.

The major difference is that maintenance does not produce any new capital asset and the measure of its effectiveness lies in the continuing ability of the building to earn a satisfactory rate of return on the total cost, i.e., initial and subsequent maintenance costs.

The aim should be to optimise the total costs as illustrated in Fig. 2.16. The direct costs represent the estimated expenditure on maintenance and the indirect costs the additional costs incurred through lack of maintenance, e.g. extra labour costs, reduced sales, etc.

It is assumed that the direct costs are in respect of work which is necessary and which is efficiently carried out. Ineffective or inefficient work will not alter the optimum standard but it will raise the total cost of achieving that standard. Also it is necessary to take into account the timescale of events and project the model into the future. Thus where work is deferred the indirect costs should include the future cost of executing the work discounted to present value. This is particularly important in the case of components which are subject to progressive deterioration and which have a steep cost growth.

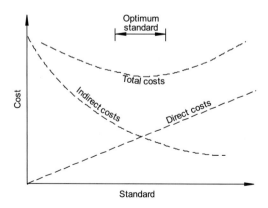

Fig. 2.16   Direct and indirect costs related to maintenance standard.

The model gives only the total cost and it would still be possible to direct resources to the repair of elements that have the least effect on indirect costs. It is necessary, therefore, to analyse each element in this way in order to obtain a series of suboptimum costs that together would represent the optimum solution for the building as a whole. There is, of course, a danger in rigidly applying financial controls which may result in essential work either not being done or being done inadequately with costly rectification later.

It should be borne in mind that laying down a fixed cost limit will result in a variable amount of work being carried out according to the prices charged by contractors. This will depend not only on the nature of the work and any special difficulties surrounding its execution, but also upon the market conditions prevailing in the locality. Thus, if the price of maintenance were to double, only half the work could be carried out with the possibility of serious long-term consequences.

Also, maintenance costs are determined by quite different factors from production and other activity costs. It is not difficult to conceive of large-scale work which is necessary to preserve the structural stability of the building, but which has no immediate effect on the use of the building, while some comparatively minor item could lead to a production stoppage with very serious financial consequences. Thus, a cost limit related to output would be completely inadequate in the one case and extravagant in the other.

It becomes even more difficult to fix standards in this way where there is a large element of subjectivity involved. For instance, internal painting and decorating forms a large proportion of building maintenance expenditure and yet the cycles are fixed on the basis of supposed visual acceptance. It is unlikely that redecorating an office every five years instead of every four years would have any measurable effect on labour costs, either resulting from a slower pace of working, higher labour turnover or the need to pay higher wages in compensation. One might assume that maintaining a high standard in areas open to the public would create a favourable image of the firm and thus result in indirect benefits. In such cases, the costs incurred would be of a similar nature to those incurred in advertising or public relations and would have to be judged according to similar criteria.

Where the object of the maintenance is to reduce the probability of injury or damage to the occupants or contents the expenditure would be akin to that incurred in insurance and the level would be related to the risks involved. Clearly no building can be perfectly safe in all respects and a proper balance must be achieved between the financial consequences of failure and the cost of work designed to prevent failure.

References

(1) Robertson, J.A. (1969) The planned maintenance of buildings and structures. *The Institution of Civil Engineers Proceedings Paper 7148S.*

(2) Kemp, J. (1966) *Design, Maintenance And Operation.* HMSO, London.

(3) Kelly, J. & Male, S. (1993) *Value Management In Design And Construction.* Chapman Hall, London.

(4) Chanter, B. & Swallow, P. (1996) *Building Maintenance Management.* Blackwell Science, Oxford.

(5) Bushell, R. (1990) In: *Maintenance Management – A Guide To Good Practice.* Chartered Institute of Building, Ascot.

(6) Stevens, R. (1973) Maintenance standards and costs. Paper given at *National Building Maintenance Conference*, London, 1973.

(7) Wordsworth, P. (1992) Standard Maintenance Descriptions: Part 1: Maintenance Management Practice. *Royal Institution of Chartered Surveyors Research Paper 20.* RICS, London.

(8) Gorter, S.J. (1992) The Significance Of Condition Monitoring Within The Scope Of Maintenance Control. Paper given at *Chartered Institute of Building/W70 Conference*, Rotterdam, 1992.

Chapter 3
Statutory Control

3.1 Understanding legislation

There is an increasingly large and complex body of legislation relevant to the management and maintenance of property, a knowledge of which is essential to the building maintenance manager. This is necessary for the provision of an efficient and professional management service which safeguards the legal rights and interests of building owners, occupiers and visitors. In general, the purpose of statutory legislation is to provide a regulatory framework in which identified problems in a particular area can be controlled and resolved when these are outside the scope of the normal operation of commercial and contractual interests, but which are considered to be in the wider interest of society or its individuals. As an example of this, statutory housing fitness standards protect the individual tenant from the risks and costs of legally challenging a large corporate property owner if the property is basically unfit for habitation. It follows that the requirements of statutory legislation are therefore to be considered as a minimum, rather than an optimum standard, intended to enforce a social consensus and to prevent unfair competition from unscrupulous companies or individuals. Lack of awareness of, or failure to comply with, these statutory legal requirements may result in the imposition of fines and notices prohibiting or restricting the use and occupation of the building; or even, in certain extreme cases of negligence relating to health and safety in particular, to criminal charges.

The following will give an overview of the most relevant and important pieces of legislation in force at the time of writing. However, new statutes, new regulations made under existing statutes, and new interpretations of existing regulations are constantly being formulated in what is a dynamic rather than a static legal environment. The problem for the maintenance manager therefore is how to develop a strategy to keep abreast of current relevant legislation.

The first step in such a strategy is in acquiring a basic understanding of the structure of the legislation. The second step is awareness of the major statutory provisions currently in force. The third step in the strategy is to develop methods of keeping up to date with changes in the legislation, both proposed (in order to make a forward assessment of how any new legislation is likely to impact on the management and maintenance of buildings), and introduced (when compliance may become a legal necessity). Each of these steps will now be dealt with in turn.

3.2 The structure of legislation

The legislation relevant to this section is statutory law; that is, laws passed by parliament and usually administered and enforced by the government or its agents. Statutory law should therefore be distinguished from civil law such as tort or contract which is based on case law and precedent, and which regulates the contracts between individuals or organisations. The situation can be confusing since some statutory law, such as the Landlord and Tenant Acts or the Party Wall Act are statutes which dictate how certain issues between individuals or companies are conducted. Nevertheless, what is clear is that if there is an act of parliament or statute, such as the above examples or such as the Health and Safety at Work Act, then it is the government or its designated agents to whom the building maintenance manager is answerable in law. However, if a failure to observe statutory provisions leads to loss or injury to another party, the maintenance manager may also face a separate civil case brought by the injured party of liability for negligence in addition to any statutory penalties imposed.

A statute is passed in full parliament and when the royal assent is given, becomes law. It is designated a title which includes the year of its enactment; for example the Building Act 1984.

It is enforced either by a government agency (such as the Health and Safety Executive), or delegated to a local authority (such as the fire authority, or a local government department such as planning and building control), or to a private statutory undertaker (such as the water companies). Where enforcement power is delegated in this way there is usually a right of appeal upwards to the delegating authority, which is ultimately and for matters of principle, the Law Lords. Planning appeals to the Department of the Environment are the most commonly encountered instance of this in property management.

The act itself sets out the principles of the legislation, and may itself contain the provisions to be complied with; but more often, they are the means by which regulations are introduced. For example, the Building Regulations are in force only as a provision of the Building Act 1984. Regulations may be directly introduced or amended by the relevant Secretary of State (usually after appropriate consultations with interest groups, commercial agencies, and professional bodies) without the necessity of a new or amended act being put through parliament, provided that the content of the regulations is entirely within the provisions of the original act. It is in this manner that the Building Regulations are in the process of regular updates; and that new Health and Safety legislation is introduced to comply with new European standards under the Health and Safety at Work Act 1974. Indeed, under the Health and Safety at Work Act, there are now well over one thousand *sets* of regulations which cover in detail the gamut of health and safety issues across industry and commerce.

Because it is impossible to foresee exactly how a new piece of legislation will actually work in practice, the interpretation of the regulations themselves may also be further clarified by *determinations,* which are specific decisions made either by courts, or by the delegating authority, from time to time, on particular interpretations of legislation. The most familiar determinations to building maintenance managers are those concerning specific aspects of fire safety legislation under the Fire Precautions Act 1971 or in the interpretation of Health and Safety at Work Regulations.

Furthermore, it is often erroneously supposed that the enforcement agency has the unqualified right to dictate the strict terms of compliance with statutes and regulations by means of prescriptive schedules and orders, whereas in many cases it is possible for the maintenance manager to actively participate in compliance by negotiating the terms with the enforcement agency. This obviously presupposes an accurate knowledge of the regulatory content by the maintenance manager, but can have great advantages in such areas as for example, fire safety engineering, where there may be a number of different ways of fulfilling the statutory requirements, one of which fits into the wider context of the building's use and maintenance better than the others. This welcome trend towards regulatory standards based on performance rather than prescription is best illustrated by the current structure and administration of the Building Regulations.

3.3 The major statutory provisions

There are a great many acts of parliament which relate in one way or another to building standards; some apply to buildings in general while others control the condition of specific types of building. Those which are of most concern to the maintenance manager may be classified as:

- Those which apply to the design and physical requirements of new construction, additions, alterations and works necessitated by a material change of use. These requirements are largely covered by the Planning Acts and the Building Regulations.
- Those which are of a continuing nature and apply during the occupation of the building. The most important provisions are contained in the Health and Safety at Work Act and the Regulations made thereunder.

Town and Country Planning Acts

The main provisions regarding planning are to be found in the Town and Country Planning Act 1990. The general principle is that a landowner has no right to use the land for any purpose other than its present use unless he or she first obtains permission for the change of use. The object of the statutory provisions is to control 'development' which is described in the act as 'the carrying out of building, engineering, mining or other operations in, on, over or under land or the making of any material change in the use of buildings or other land'. However, certain operations are deemed not to involve development and include works of maintenance, improvement or other alterations which affect the interior of the building but do not materially affect the external appearance. Also the act empowers the Secretary of State to make development orders which exempt the developer from obtaining planning consent for the permitted development. The Town and Country Planning (General Permitted Development) Order 1995 (SI 1995 No. 418) granted such development consent to the enlargement, improvement or other alteration of buildings according to their use class provided the development falls within the order's guidelines. Except in cases of permitted development it is necessary for planning permission to be obtained from the local planning authority which may grant it conditionally or unconditionally, or may refuse it.

The Planning (Listed Buildings and Conservation Areas) Act 1990 also provides for the Secretary of State to compile lists of all buildings which are of historical or architectural interest, known as listed buildings. Buildings are selected on the grounds that they are good examples of a particular architectural style or piece of planning, that they incorporate some technological innovation, that they have associations with well-known characters or events, or for their group values, e.g. squares and terraces. In practice any building dating before 1700 is automatically listed and most of those built between 1700 and 1840. Listed buildings are graded as follows:

- *Grade I listed.* 6000 UK properties of major historic interest. This group includes monuments such as Stonehenge, and many major cathedrals. Any repairs and alterations are very strictly controlled.
- *Grade II* listed.* 23 000 properties of significant historic interest, including many country and town houses.
- *Grade III listed.* 400 000 properties of special interest, that is, with some particular notable feature, or part of a conservation area.

Listed building consent must be obtained from the local planning authority for demolition, alteration or extension of any listed building in any manner which would affect its character. Consent is therefore necessary even in the case of such minor alterations as removing the glazing bars from Georgian windows. It is interesting to note that a local authority can now prevent an owner from destroying a listed building through deliberate neglect by serving on the owner a repairs notice specifying the work reasonably necessary to preserve the building. In the event of the necessary work not being carried out the authority or the Secretary of State may make a compulsory purchase order and pay only minimum compensation. Also where it appears to the local authority that the owner has demolished or altered part of the building in contravention of the act they may serve on the owner a listed building enforcement notice requiring him or her to restore the building to its former state.

Further controls were introduced by the Town and Country Amenities Act 1974 which applied listed building controls to all buildings within conservation areas. Thus anyone wishing to demolish a building within a conservation area must apply for listed building consent to the local planning authority either separately or as part of an application for planning permission to redevelop the site. In such cases special regard should be paid to such matters as bulk, height, materials, colour, vertical and horizontal emphasis and grain of design. The Secretary of State has power to make grants for the maintenance and repair of buildings of 'outstanding' architectural or historic interest and for the preservation or enhancement of character or appearance of conservation areas.

One further piece of planning legislation which may affect maintenance managers is the Town and Country Planning (Assessment of Environmental Effects) Regulations 1988 (SI 1988 No. 1199), whereby the environmental impact of any proposed development or change of use must be assessed as part of the application process.

The Building Act 1984

This act consolidates most of the primary legislation relating to the standards of building construction. The main provisions in the context of maintenance and improvement work are:

Part I Building Regulations

Section 1 gives the Secretary of State the power to make regulations with respect to the design and construction of buildings and the provision of services, fittings and equipment in or in connection with buildings for the purposes of:

- securing the health, safety, welfare and convenience of persons in or about buildings and of others who may be affected by buildings
- furthering the conservation of fuel and power
- preventing waste, undue consumption, misuse and contamination of water.

It should be noted that these are the only criteria by which the Building Regulations can be enforced. Poor workmanship or inappropriate specification which does not affect compliance with these three criteria, even when it may result in economic loss to the owner (such as in the case of subsidence) are not subject to the provisions (cf. *Murphy* v. *London Borough of Brentwood*). Building regulation approval is therefore not a complete substitute for quality inspections of building works by the clerk of works or supervising surveyor.

Section 2 has not been enacted to date but it provides that Building Regulations may impose continuing requirements on the owners and occupiers of buildings including those which were not at the time of their erection subject to regulation.

Section 6 provides for the issuing of 'approved documents' by the Secretary of State, for the purpose of providing practical guidance on the provisions of the Building Regulations. Whilst these are not mandatory, the burden of proof as to compliance rests with the applicant if they are not used.

Section 8 gives the Secretary of State power to relax Building Regulations after consultation with the local authority or for this power to be exercised by the authority where the regulations so provide if it is considered that the regulation would be unreasonable in the particular circumstances.

Sections 19 and 20 relate to short-lived materials and the use of materials unsuitable for permanent buildings and provide that a local authority may pass plans showing the use of such materials subject to fixing a period on the expiration of which the building must be removed, or imposing conditions on the use of the building.

Part II Supervision of building work, etc., other than by local authorities

This part provides for the optional system of supervision of plans and work by an approved inspector and lays down the procedures to be followed in respect of giving the initial notice to the local authority and the final certificate. It is chiefly used for newbuild private housing as an alternative to local authority control, but private inspectors are increasingly being used in the commercial sector.

Part III Other provisions about buildings

This part gives local authorities the power to require the owners or occupiers of buildings to provide certain facilities or to remedy certain deficiencies. For this purpose the authority may give 'notices' requiring:

(1) Provision of a satisfactory drainage system or the execution of such work as may be necessary for renewing, repairing or cleansing drains and appliances (S59).
(2) Discontinuance of the use of rainwater pipes for the purpose of conveying soil or drainage from a sanitary convenience (S60).
(3) Provision of sufficient closet accommodation (S64).
(4) Provision of sufficient and satisfactory sanitary conveniences in a building used as a workplace (S65).
(5) Replacement of earth closets where a sufficient water supply and sewer are available (S66).
(6) Provision of a wholesome water supply in occupied houses (S69).
(7) Provision of sufficient and suitable accommodation for food storage in a house or part of a building occupied as a separate dwelling (S70).

(8) Provision of such means of ingress or egress to the building as the authority after consultation with the fire authority deem necessary (S71).

(9) Provision of satisfactory means of escape from fire in the case of flats and other designated buildings which exceed two storeys in height and in which the floor of any upper storey is more than twenty feet above ground level (S72).

(10) Raising of chimneys where a building is erected or raised to a greater height than an adjoining building (S73).

(11) Alteration or closing of cellars and rooms below subsoil water level if constructed without the consent of the local authority (S74).

(12) Intention of the local authority to remedy the defective state of a building where the serving of an 'abatement notice' under Sections 93 to 96 of the Public Health Act 1936, would result in unreasonable delay (S76).

(13) Intention of the local authority to take emergency measures where a building is in such a state or is used to carry such loads as to be dangerous (S78; see below).

(14) The repair, restoration or demolition of a building which is by its ruinous or dilapidated condition seriously detrimental to the amenities of the neighbourhood (S79; see below).

(15) The demolition of a building stating work to be done to safeguard adjoining buildings, e.g. shoring, weatherproofing exposed surfaces, making good any damage caused by the demolition, etc. (S81).

(16) Paving and drainage of yards and passages (584).

Dangerous buildings and structures

Sections 77 and 78 deal with dangerous buildings and structures. If the local authority deems a building to be in danger of structural collapse, it will first issue a Notice (S77) requiring the building to be made safe, or overloading to be discontinued. If within the specified time the owner does not respond to this Notice, the local authority may, on application to the magistrates court for an Order, force its evacuation, and may carry out such works as are necessary to secure the building so that it does not endanger the public (S78). The cost of any works carried out by the authority can be recharged to the building owner by means of a land charge on the land registry entry for the premises.

Building Regulations 1985

The Regulations are made under the Building Act 1984 and apply in England and Wales to building works and certain changes of use of an existing building. Scotland and Northern Ireland have separate sets of regulations.

Building work is defined in Regulation 3 as:

- the erection or extension of a building
- the material alteration of a building

- the provision, extension or material alteration of a controlled service or fitting
- work required on a material change of use.

Building is defined as any permanent or temporary building but not any other kind of structure or erection, and includes part of a building. Certain small buildings and extensions as well as certain buildings used for special purposes are exempt from the regulations.

The regulations say nothing about the point at which repair becomes subject to control. Normally repair whether it involves replacement or making good would not be controlled, but where, for example, a whole building has been seriously damaged the repairs could be so extensive that the local authority could reasonably treat the work as the erection of a new building and apply the regulations. Most day-to-day maintenance works therefore fall outside the scope of the Building Act and Building Regulations. However, those involving demolitions or alterations, particularly to the structure and specifically to replacing a traditional roof covering, are subject to the regulations.

The regulations provide for two systems of control, i.e., supervision by the local authority or by a private approved inspector. In the former case the person intending to carry out building work has the option of either depositing full plans (mandatory in the case of offices and shops) or of giving a building notice containing much less detail. The advantage of giving full plans is that if they have been passed by the local authority and the work complies with them the local authority may not subsequently require the work to be taken down or altered. Also where full plans are accompanied by a certificate given by an approved person to the effect that the plans show compliance with certain requirements, e.g. structural stability or energy conservation, the local authority cannot reject the plans on the grounds of noncompliance with the requirements to which the certificate relates.

Applications for building control approval to the local authority can either be by Building Notice or by Plan Deposition.

Building Notice is a simple procedure suitable for small alterations which do not affect the current provisions for means of escape in case of fire. A notice on a prescribed form is submitted to the local authority with a simple text description of the proposals. A site inspection by the local authority building control office will be used to determine any statutory requirements. A small fee is payable on a fixed scale, to the local authority.

Full Plans are submitted to the local authority for bigger schemes, or those involving means of escape. A fee both for the plan inspection, and subsequent statutory site visits by the building control officer, is payable according to a fixed scale. Notwithstanding compliance with the Building Regulations a local authority may reject the plans of a proposed building or extension if they fail to show:

- proper drainage facilities (S21)
- satisfactory means of access for removal of refuse to a street (S23)
- adequate fire exits (524)
- provision of an adequate supply of wholesome water to a house (525)
- sufficient closet accommodation (S26)
- provision of a bathroom containing a fixed bath or shower and hot and cold water supply (S27)
- suitable accommodation for food storage to dwellings (S28)
- freedom of building site from offensive material (S29).

Persons contravening the Building Regulations are liable to be fined and the local authority may issue a 'Section 36 Notice' requiring the owner to either pull down the work or, if he or she so elects, to effect such alterations as may be necessary to secure compliance. An appeal against such a notice may be made under Section 40.

If it is intended that the work should be supervised by a private approved inspector the person intending to carry out the work and the inspector should jointly give to the local authority an 'initial notice'. The local authority must accept or reject the initial notice within ten working days but once they have accepted it their powers to enforce the regulations are suspended. Unless the work consists of the alteration of a one or two-storey house the approved inspector must be independent of the designer or builder. Unlike the local authority an approved inspector has no direct power to enforce the regulations but must inform the person responsible for the work that he or she believes the work is in contravention of the regulations and if the alleged contravention is not remedied within three months he or she is obliged to cancel the initial notice. In that event the local authority becomes responsible for supervision and may request sufficient plans of the work and may require the opening up of work to ascertain if any uncertified work contravenes the regulations.

The Approved Documents are published separately and give practical guidance on the ways of meeting the regulations. Other means of satisfying the regulations may be adopted bearing in mind that in relation to Parts A–N nothing needs to be done beyond that which is necessary to secure reasonable standards of health and safety. Approved Documents may give guidance in the form of technical solutions or acceptable levels of performance. If there is no technical solution which is suitable in the particular circumstances an alternative approach may be adopted based on the relevant recommendations of a British Standard. It should be noted that British Standards and British Board of Agrément Certificates often cover serviceability or recommendations for good practice which go beyond the requirements of the regulations.

The Fire Precautions Act 1971

This act (together with a number of subsequent related acts) forms the statutory basis for fire protection and means of escape requirements in all buildings other than single family houses, and is of primary importance to the building maintenance manager. The object of the act is to ensure the provision of adequate means of escape and related fire precautions; and, unlike the Building Regulations, compliance with the act is a *continuing requirement* for the occupation of designated premises. The key feature of this continuing requirement is the Fire Certificate for the designated premises, which defines the approved fire precautions for that building.

Premises affected

Premises are brought within the scope of the act either following the issue of a designation order or when the Secretary of State makes regulations applying fire certificate provisions to the premises or when the fire authority issues a notice relating to a particular building.

The designation order must fall within the following classes of use in order to define a type of building for which a fire certificate is required:

- sleeping accommodation
- institution providing treatment or care
- entertainment, recreation or instruction, or for the purpose of a club, society or association
- teaching, training or research
- any other purpose involving access to the premises by the public whether on payment or not
- any place where people work.

The first designation order was in respect of hotels and boarding houses and came into effect on 1 June 1972. This brought under control premises used for the business of a hotel or boarding house where:

- sleeping accommodation is provided for more than six persons being staff or guests, or
- some sleeping accommodation is provided for staff or guests above the first floor, or
- some sleeping accommodation is provided for staff or guests below the ground floor.

The 1971 Fire Precautions Act has also been amended by the Health and Safety at Work, etc. Act 1974, which designated factories, offices, shops and railway premises for which fire certificates are necessary. Fire certificates are also required for certain premises in which explosive, inflammable or other hazardous materials are stored or in which hazardous processes are carried out.

Subsequent related acts and orders which broaden the scope of the original act include the Fire Safety and Safety of Places of Sport Act 1987, and the NHS and Community Care Act 1990, which removed the Crown immunity previously enjoyed by health authorities in respect of hospitals and healthcare buildings. More recently, the Fire Precautions (Workplace) Regulations 1997 (SI 1997 No. 1840) extended the provisions to include any place where people work, though this does not always necessitate obtaining a Fire Certificate in the case of smaller premises employing less than 20 persons, or 10 persons other than on the ground floor.

Observance of the Act

The primary responsibility lies with the *occupier* of the appropriate part of the building although the fire authority can impose requirements on others. For the smaller premises referred to above, there is a requirement that reasonable precautions are taken to ensure means of escape and the means of first-aid firefighting are met. This is usually done by undertaking an assessment of the particular risks associated with the premises and their use.

The Fire Certificate

For all other designated premises, application must be made to the fire authority for a fire certificate when the building starts to be occupied. Following inspection of the premises the fire authority will issue a certificate specifying:

- the use or uses of the premises which it covers
- the means of escape in case of fire
- the method of ensuring that means of escape can be safely and effectively used at all material times
- the means of fighting fire for use by persons in the building
- the means of giving warning in case of fire.

Additionally, the fire certificate may impose requirements relating to:

- maintenance of the means of escape and keeping them free from obstruction
- maintenance of other fire precautions specified in the certificate
- training of staff and keeping records
- limitation on the number of people who may be in the building at any one time
- any other relevant fire precautions.

If the use or occupancy pattern of the building subsequently change, or if any alterations affecting the means of escape or other matters designated in the fire certificate, the fire authority must be informed. Further requirements may then be imposed, and the fire certificate amended if necessary. The fire authority also has the right to make spot checks on premises to determine whether the designated fire precautions measures are being complied with. This includes an adequate inspection and maintenance regime for firefighting equipment such as extinguishers and alarms, for which service records should be kept.

Breaches of fire certificate conditions may in serious cases result in the issue of a Section 10 Order closing the building, and may also result in criminal charges being made against the occupier. Less serious or reckless breaches of the requirements will result in enforcement orders and fines.

It is also a requirement to keep available an up-to-date copy of the Fire Certificate at the designated premises.

The Health and Safety at Work Act 1974

This act provides a comprehensive and integrated system of law governing the health, safety and welfare of workpeople and the health and safety of members of the public who are affected by work activities. The Act itself is basically an enabling act which sets out the general principles of health and safety management, and which provides the framework for the associated regulations in force under this act. There are a considerable number of sets of regulations, and more are being introduced over time, particularly in response to EC health and safety directives. Only the more important regulations are outlined in this section.

The act is in three parts: Parts I and II, relating to safety and health, are administered by the Department of Employment; Part III, which extended the scope of the Building Regulations, has been incorporated in the Building Act 1984.

Part I

The provisions of this part impose general duties in relation to health and safety on the following persons:

(1) Employers are to ensure, so far as is reasonably practicable, the health, safety and welfare at work of their employees.
(2) Both employers and self-employed persons are to conduct their undertakings in such a way as to ensure, so far as is reasonably practicable, that persons not in their employment are not exposed to risks to their health or safety.
(3) Persons in control of nondomestic premises which are used by persons not in their employment either as a place of work or as a place where they may use any plant or substance provided for their use must ensure, so far as is reasonably practicable, that the premises and any plant or substance therein is, when properly used, safe and without risk to health. In addition there is a duty to use the best practicable means to prevent the emission into the atmosphere of noxious or offensive substances.
(4) Persons who design, manufacture, import, install or supply any article for use at work must ensure that the article is safe and without risk to health when properly used, and are required to carry out such tests as are neces-

sary and to take such steps as are reasonably practicable to ensure that adequate information is available concerning the correct use of the article.

(5) Employees must take reasonable care for the health and safety of themselves and of other persons who may be affected by their actions and shall
not interfere intentionally with or misuse anything provided in the interests
of health, safety or welfare.

Regulations may be made describing the circumstances in which safety representatives may be appointed or elected from among the employees to represent
them in consultations about health and safety. In certain circumstances the employer may have a duty to establish a safety committee to keep the health and
safety measures under constant review.

This part of the act also provided for the creation of the Health and Safety
Commission and the Health and Safety Executive. The Commission has general
responsibility for the work of its Executive which is the main body responsible
for enforcing the statutory requirements on health and safety. Enforcement powers are also given to local authorities and certain other bodies.

The Health and Safety Commission consists of a chairman with a minimum
of six and a maximum of nine other members appointed after consultation with
employer and employee organisations, local authorities and other organisations. It is responsible for the continuous task of preparing proposals for the revision and extension of statutory provisions on health and safety and for making
appropriate arrangements for research and training.

The Health and Safety Executive consists of three persons appointed by the
Commission with staff drawn from the previous health and safety inspectorates
for factories, mines and quarries, alkali works, etc. Thus, if an inspector is of
the opinion that a person is contravening a statutory provision the inspector may
issue an 'improvement notice' requiring the person to remedy the contravention.
If the inspector considers that the activity involves a risk of serious personal
injury he or she may issue a 'prohibition notice' requiring the cessation of the
activity.

Part II
This part relates to matters concerning the safeguarding and improvement of the
health of employed persons.

Observance of the Act
The maintenance manager has much of the responsibility for protecting the interests of the occupier in providing safe working conditions, safety of access
and health and welfare facilities. It should be noted that the act places a responsibility on the person in control of the premises to ensure that not only are the
premises themselves safe but also any plant or equipment within the building.
This imposes a duty on the person in charge to ensure that the plant and work
methods used by outside contractors are not prejudicial to the health and safety

of employees or to others who have rightful access to the building. Thus the maintenance manager could incur a personal liability for injuries caused by such contractors. Also, there is likely to be an increase in the amount of work which has to be done in response to requests from safety representatives.

The safety duties imposed by the act may be summarised as follows:

(1) to make and keep all workplaces safe s2(2)(d) and (e)
(2) to ensure the safety of machines and materials s2(2)(a) and (b)
(3) to plan and use safe working systems s2(2)(a) and (e)
(4) to train, inform and direct employees s2(2)(c) and (3)
(5) to receive and consider employee views s2(6) and (7)
(6) to ensure the safety of subcontractors and visitors s4(1) and (2)
(7) to protect the wellbeing of co-tenants, neighbours and the public at large s3(i)
(8) to test and supply in safe condition any goods for use by employees or others s6(1), (2) and (3)
(9) to effectively safeguard all employees wherever they may work s2(1)
(10) to set out in writing the arrangements and organisation to achieve the above.

Enforcing measures include:

- criminal prosecution of the employer, controllers of premises, designers, manufacturers and suppliers of commodities used at work
- prohibition or improvement notices
- industrial tribunal complaint in respect of (5)
- civil claims based on alleged negligence or breach of statutory duty.

It should be noted that where a statutory provision requires certain things to be done it is no excuse for the employer to claim that it was not practicable unless the provision states otherwise. Also it is not possible to escape the liability by delegation to others. The only exception is where the employer has delegated the duty to carry out a statutory provision to someone for whose protection the provision was imposed and he or she is injured through their own negligence, i.e., the employer may claim contributory negligence. This doctrine of 'best practicable means' requires managers to demonstrate that in the event of an accident, they had taken appropriate steps to minimise such a risk. This underlines the need for careful record keeping of maintenance and servicing schedules, risk assessments, and internal health and safety documentation.

There are a number of regulations in force under the act which have far reaching implications for the maintenance manager, the most important of which are outlined below.

Control of Substances Hazardous to Health Regulations 1999 (COSHH)

These regulations relate to the safe use of materials used or generated in the workplace, which include those used in maintenance works. It requires an assessment of the risks of using the material (for example, of maximum exposure limits to mortar dust as a trigger for asthma), and a methodology for its safe use in the circumstances. Whilst most contractors and suppliers now routinely provide such assessments in relation to their own workforce, it must be noted that in the case of maintenance works in occupied premises, it is essential to ensure that the risks to all occupiers are assessed, and screening, dust extraction, and warning notices provided as appropriate.

Workplace (Health Safety and Welfare) Regulations 1992

These regulations require the minimisation of certain types of risk, notably section 13, which requires occupants to be excluded from potentially dangerous areas, for example where maintenance works are in progress. Other requirements, which replace previous standards in force under the Offices Shops and Railway Premises Act 1963 and the Factories Act 1961, include:

 (1) workplace equipment maintenance
 (2) ventilation, lighting, and temperature
 (3) workspace, workstations and seating
 (4) condition of floors and circulation routes
 (5) falls, trips, and falling objects
 (6) replacing annealed glazing with safety glazing or protective films
 (7) safe operation of doors, gates, escalators and moving walkways
 (8) provision of adequate sanitary conveniences and washing facilities
 (9) cleanliness generally, and storage and disposal of waste materials
(10) provision of drinking water
(11) accommodation for clothing and changing facilities
(12) rest and meals facilities
(13) facilities for pregnant women and nursing mothers
(14) prevention of tobacco smoke affecting others.

These regulations therefore have a significant impact on both the standards and execution of maintenance works, and are therefore of great importance to the maintenance manager.

The Control of Asbestos at Work Regulations 1987

(Amended 1992 and 1998.) These regulations place strict controls on asbestos removal in buildings, and require the Health and Safety Executive to be notified of any such works. These regulations are to be replaced by the Control of Asbestos at Work Regulations 2001, which will require in addition, a formal risk assessment to be made for all buildings which contain or may contain asbestos, in accordance with an approved Code of Practice, which is in preparation at the time of publishing.

Provision and Use of Work Equipment Regulations 1992

This imposes a requirement for the regular inspection and maintenance of all workplace equipment, including electrical appliances. It also requires that equipment must be capable of being maintained safely, which may include a requirement for training operatives.

Other regulations under the Health and Safety at Work Act 1974 relevant to maintenance works include the Construction (Health Safety and Welfare) Regulations 1992 and the Construction (Design and Management) Regulations 1994 (CDM Regulations). These lay down specific requirements for risk assessments and the preparation of a health and safety plan for certain works which may include larger maintenance projects (those continuing for more than 30 working days or where more than 500 person-days are to be spent). In the case of the CDM regulations there is a requirement for the preparation of a Health and Safety File relating to subsequent use and maintenance, which should be handed to the occupier on completion of the works.

The Defective Premises Act 1972 and the Occupiers' Liability Acts 1957 and 1984

Defective Premises Act 1972

The purpose of the act is to impose duties in connection with the provision of dwellings and to amend the law as to liability for injury or damage caused to persons through defects in the state of premises.

Duty to build dwellings properly

A person taking on work for or in connection with the provision of a dwelling (whether by the erection or conversion or enlargement of a building) owes a duty to see that the work taken on is done in a professional manner with proper materials and that the dwelling will be fit for habitation when completed. The duty is owed to the person for whom the dwelling is provided and to every person who subsequently acquires an interest in the dwelling. Thus the duty is imposed upon builders, subcontractors, architects and surveyors who take on work of the type described and upon developers and others who arrange for builders to take on such work. As this is a statutory duty it cannot be excluded by terms in the building contract. Further, the duty is owed to persons who acquire an interest in the dwelling and were not parties to the original contract.

Where, however, a person takes on work of this nature on the condition that the work is be done in accordance with instructions given by or on behalf of another person he or she will be relieved of this obligation to the extent that the instructions have been complied with and provided that he or she does not owe a duty to warn the other person of defects in the instructions and has failed to do so. A person shall not be treated as having given instructions merely because of having agreed to the work being done in a specified manner with specified materials to a specific design. For example, where a builder agrees to provide a dwelling in accordance with plans and specification of his own he does not escape the statutory duty by merely adhering to the agreed plans and specification. No action under this provision may be brought by a person having or acquiring an interest in a dwelling where rights in respect of defects are conferred by an approved scheme.

Duty with respect to work done on premises not abated by disposal
The act provides that where work of construction, repair or demolition or any other work is done on or in relation to the premises, any duty of care owed in consequence of the doing of the work to persons who might reasonably be expected to be affected by defects in the state of the premises created by the doing of the work, is not to be abated by the subsequent disposal of the premises by the person who owed the duty.

A person who does work to any premises is under a duty at common law to take reasonable care for the safety of others who might reasonably be expected to be affected by defects in the state of the premises arising from such work. Hitherto, however, his or her liability was extinguished on disposal of the premises. A purchaser or tenant taking premises in such circumstances was affected by the doctrine of *caveat emptor* and could not claim for negligence against the vendor or lessor. The act now removes this special immunity of vendors and lessors from liability for negligence.

These provisions do not apply:

(1) Where the disposal was a letting and the 'tenancy' commenced before 1 January 1974.
(2) In the case of premises disposed of in any other way, if the disposal was completed or a contract for the disposal was entered into before 1 January 1974.
(3) If the relevant transaction disposing of the premises is entered into in pursuance of an enforceable option by which the consideration for the disposal was fixed before 1 January 1974.

Landlord's duty of care
Where premises are let under a 'tenancy' which puts on the property owner an obligation to the tenant for the maintenance or repair of the premises, the property owner owes to *all persons* who might reasonably be expected to be affected by defects in the state of the premises, a duty to take such care as is reasonable in the circumstances to see that they are reasonably safe from 'personal injury' or from damage to their property caused by the 'relevant defect'. The duty is owed if the property owner knows (whether as a result of being notified by the tenant

or otherwise) or if he or she ought in all the circumstances to have known of the relevant defect. The term 'relevant defect' means a defect in the state of the premises existing at the material time, i.e., 1 January 1974 for tenancies entered into before that date or the date when the tenancy commenced, and arising from or continuing because of a failure by the property owner to carry out the repairs that are obligatory.

This provision replaces Section 4 of the Occupiers' Liability Act 1957, and extends to all those who might reasonably be expected to be affected by the defect (e.g. a passer-by on the highway or neighbours in their gardens), the duty of care previously owed merely to 'visitors' to the premises.

Where the premises are let under a 'tenancy' which expressly or implicitly gives a property owner a right to enter premises to carry out maintenance and repair work then, as from the time when he or she first is, or by notice can put him or herself, in a position to exercise the right , then the property owner is to be treated for the purposes of this provision as if he or she were under an obligation to the tenant for that description of maintenance or repair of the premises.

The Occupiers' Liability Act 1957 and the Occupiers' Liability Act 1984

These acts lay down the 'duty of care' owed by the occupier of a building to those visiting the building, whether authorised or not. If, through lack of maintenance or deliberate act, the visitor is put in danger of accident, the occupier is liable to prosecution. Penalties include fines and imprisonment in serious cases.

The Environmental Protection Act 1990

This act replaces and consolidates previous public health acts relating to statutory nuisances (Part III). Section 79 defines classes of statutory nuisance: ('prejudicial to health or a nuisance').

- premises in a condition to be prejudicial to health or a nuisance
- smoke emitted from premises
- fumes, gases, dust, steam
- noise
- animals
- dirt or accumulations.

The local authority can issue a Section 80 Abatement Notice which specifies what must be done to abate the nuisance and prevent its recurrence, within specified time limits.

Waste disposal arrangements also come within the purview of this legislation, which imposes a duty of care on building owners to store and dispose of waste in the correctly defined manner.

The Housing Acts

Standards relating to residential premises are laid down in various Housing Acts and relate to:

- the condition of the fabric
- the equipment and services
- the quality of the surrounding environment
- repair, stability, dampness, natural lighting and natural ventilation
- sanitary fittings, hot and cold water supply, drainage, cooking facilities, artificial lighting, heating installations
- space for activities and circulation, privacy in houses in multiple occupation
- air pollution, noise level, open space, traffic conditions

Basic statutory standards of fitness for rented housing are required in the Housing Act 1985, amended by the Local Government and Housing Act 1989, which introduced a new Section 604 into the former act. This Section lays down the minimum standards of fitness as follows:

To be fit for occupation, a dwelling house should:

- be free from serious disrepair
- be structurally stable
- be free from dampness prejudicial to the health of the occupants
- have adequate provision for lighting, heating and ventilation
- have an adequate piped supply of wholesome water
- have an effective system for the drainage of foul, waste, and surface water
- have a suitably located WC for the exclusive use of the occupants
- have a bath or shower, and a wash hand basin, with hot and cold water
- have satisfactory facilities for the preparation and cooking of food including a sink with hot and cold water.

Additionally, in the case of flats, there are requirements relating to the building structure and common areas, that:

- the building or part is structurally stable
- it is free from serious disrepair
- it is free from dampness prejudicial to the occupants' health
- it has adequate provision for ventilation
- it has an effective system for the drainage of foul, surface, and waste water.

In more detail, the fitness standards are as follows:

(1) *Repair.* The state of repair should not be a threat to the health of, or seriously inconvenience, the occupiers. The internal decorative condition is not taken into account. Unfortunately it is difficult to show a direct causal relationship between disrepair and ill health and the judgement must be purely subjective and based on what is considered to be socially acceptable.

(2) *Stability.* There should be no indications of further movement which may constitute a threat to the occupants. Observation over a period of time is necessary to determine whether a crack or other defect is indicative of progressive collapse. However, the danger might be inherent in the design and there might be no prior visual signs that the building is likely to collapse.

(3) *Freedom from damp.* Dampness should not be so extensive as to be a threat to health. Dampness is a consequence of lack of repair and the same subjective judgement would have to be made as to whether the extent of the dampness was tolerable.

(4) *Natural lighting.* There should be sufficient light for normal activities under good weather conditions without the use of artificial light. No absolute standard is laid down but in the case of *Semon & Co* v. *Corporation of Bradford* (1922) it was decided that the critical level of daylight that separated what was considered to be inadequate from what was considered to be adequate was a sky factor of 0–2%. More recent cases suggest that a higher standard would be required to meet presentday needs.

(5) *Heating and Ventilation.* There should be adequate space and water heating facilities in the dwelling. There should also be adequate ventilation of all habitable rooms and working kitchens to the open air. The Building Regulations require the opening parts of windows to be not less than one-twentieth of the floor area and this may be used as the basis for assessing the adequacy of the ventilation.

(6) *Water supply.* There should be an adequate and wholesome supply of water within the house.

(7) *Drainage and sanitary conveniences.* There should be a readily accessible WC in a properly lighted and ventilated compartment.

(8) *Washing facilities.* There should be at least one bath or shower, and one wash hand basin, each with a supply of hot and cold water and proper drainage disposal.

(9) *Facilities for preparing and cooking food and for the disposal of waste water.* There should be a sink with an impervious working surface, a piped water supply and cooking appliance.

The local authority may inspect premises where a complaint has been made, and serve notices requiring the owner of the premises to rectify the shortcomings specified in the notice within a specified period of time, failing which it may carry out the works in default and recharge the owner. There are appeal provisions if the owner is aggrieved by the notice. These procedural requirements are not in force in relation to dwellings owned by the local authority itself, though it is expected to comply with the substantive content of the minimum standards laid down.

Leasehold Reform, Housing and Urban Development Act 1993

This act grants the 'right to buy' to the occupiers of rented housing owned by municipal or social property owners. It also contains provisions for the 'right to repair' of occupiers who may be aggrieved by a long response time or poor repairs service provided by the property owner. Under this legislation, the tenants have a right to action their own repairs to a value of £250, and to claim back this sum from the property owner on submitting proof of expenditure.

The Disability Discrimination Act 1995

Under this legislation, building owners and employers are obliged to 'make reasonable adjustments if their employment arrangements or premises place disabled persons at a substantial disadvantage compared with non-disabled persons'. It thus requires employers and building owners to provide access for people with disabilities to all parts of their existing premises where people work; or if this cannot be achieved to the full standard because of the nature of the building, to make appropriate management and personnel arrangements to prevent the people with disabilities from being discriminated against on the grounds of disability. There are obviously possible conflicts in the provision of access under this act, and the provision of adequate means of escape for this class of occupant required under the Fire Precautions Act. However, at the time of writing, the full implications of this act on the practice of building maintenance management have yet to be gauged by the establishment of precedents and legal yardsticks. It seems likely that the broad nature of the requirement will force a rethink of internal building layouts, means of escape, and the provision of sanitary facilities in a wide range of commercial buildings.

The Party Wall Act 1996

This act, which came into effect from July 1997, extends to the whole of England and Wales the statutory party wall provisions previously only in force in Inner London. For work that involves building operations (including maintenance) on or near a party wall or structure, it is necessary for the building owner to agree a Party Wall Award with the adjoining owner. The award covers such matters as existing condition of party structures before works commence, disruptions and reinstatements, and the manner and timescale over which the work will be executed, including access arrangements. Each party appoints either one surveyor to act for them both by agreement; or each may appoint their own surveyor. Failure to agree the award leads to mandatory arbitration by a third, independent surveyor. All costs are usually borne by the building owner, except where the adjoining owner stands to benefit directly, for example, in receiving a right to support from a new party wall structure which the adjoining owner may subsequently utilise in his or her own alteration and extension works.

The Access Onto Neighbouring Land Act 1994 is a linked piece of legislation which can be utilised in order to carry out essential maintenance to a building which necessarily involves access onto a neighbouring curtilage, where the neighbour refuses to grant access voluntarily. On application to a magistrates court, an order may be issued which specifies the manner and the timescale of such access, which the adjoining owner is bound to respect.

3.4 Keeping up to date with legislative changes

The above section gives an overview of the principal acts and regulations in force at the time of writing. However this is likely to change slowly over time as new acts and regulations are brought into force, and others repealed or amended. There are several overlapping strategies and methods by which the building maintenance manager can keep abreast of such changes.

Subscription to an automatically updated reference publication

There are several such publications relevant to property management on the market; for example Croner's Premises Management[1], Croner's Health and Safety[2], or Knight's Building Control Law[3]. Updates, usually quarterly, are sent to subscribing organisations or individuals either as hard copy or electronically by CD-ROM or disc, or both. New statutes, regulations, and determinations may be outlined or covered in detail depending on importance and relevance.

Publications by statutory bodies

Many of the statutory bodies publish guidance on their legislation, and will often retain organisations and individuals on a mailing list to advise of any new publications. Her Majesty's Stationery Office (HMSO) also publishes the full text of acts and statutory instruments on its website[4]. The Health and Safety Executive's range of publications is a good example.

Professional and trade journals and magazines

Frequently, articles on new legislation and its impact are to be found in professional journals (for example, *Chartered Surveyor Monthly*), and other similar publications. Whilst these may be more sporadic and lack the detail of the updated reference publications, they often highlight where particular challenges may lie for the practitioner. They should be seen as a supplementary rather than the primary source of legal updates.

Courses and conferences

Both new and existing legislation gives rise to professional conferences and updates, run by professional institutions, private organisations, and universities and colleges teaching relevant subjects such as building surveying, building maintenance management, or facilities and property management. Many of these may be designated as continuing professional development (CPD) courses, required by many professional organisations as a condition of continuing membership. Prices and quality of these courses can vary widely, but at their best, they provide both good information and a forum in which the impact of new legislation can be discussed with fellow professionals.

In a building maintenance management organisation, a combination of the above methods of legal updating may be employed depending on size and circumstance. However, it is essential to have some strategy in this respect, including an adequate budget for subscriptions, CPDs and conferences. Whilst it may be prudent and cost-effective to have one small group or individual specialising in maintaining and administering this current knowledge of changing legislation, it is however essential that there are mechanisms within the organisation by which this knowledge is held, and communicated to all relevant staff, by means of information memoranda, circulation lists for publications, internal short courses, and the maintenance of an up-to-date reference library.

References

(1) Croner (updated quarterly) *Premises Management* Croner Publications, London
(2) Croner (updated quarterly) *Health and Safety Management* Croner Publications, London
(3) Knight's *Building Control Law* (updated quarterly) Knight's Publishers, London
(4) Department of the Environment, Transport and the Regions website http://www.detr.gov.uk

Chapter 4

Maintenance Planning

4.1 The scope and nature of planning

The Pocket Oxford Dictionary defines a plan as 'a method or procedure for doing something: design, scheme, or intention'. In building maintenance management, the term has both a narrow and a wide definition. The narrow definition is related to preventive maintenance in the form of planned maintenance programmes, discussed in Chapter 1. The wide definition is of planning as an essential management tool for controlling all aspects of a building maintenance management operation, and will include:

- planned preventive maintenance programmes (the 'narrow' definition).
- planned levels of expenditure on day-to-day and reactive maintenance
- disaster planning
- planned strategies of asset management in conjunction with a business plan or corporate plan.

The Royal Institution of Chartered Surveyors' Guidance Note on Planned Building Maintenance[1] follows the wider definition: 'It will be seen ... that the planning of building maintenance is far broader than the planning of preventive maintenance.' The Guidance Note defines the broad remit of maintenance planning under five headings:

- determining the policy for maintenance
- deciding and preparing maintenance programmes and obtaining funds for them
- getting the work done
- controlling progress of work and budget expenditure
- monitoring the effectiveness of the programme.

It is important to note the difference between maintenance planning and programming. *Planning* embraces the whole process of maintenance management as detailed in the five categories above. *Programming* relates to scheduling the manner in which maintenance works will be carried out, defined in the Guidance Note as:

- planned maintenance programmes (in the narrow definition)
- day-to-day or breakdown maintenance
- minor new works
- refurbishment.

The nature of planning in building maintenance involves determining systems and sequences of operation. Critically, this must necessarily involve a level of prediction. These predictions relate not only to the state of the building stock and its degree and manner of deterioration, but also to the future policy of the owners or users of the buildings; economic predictions about the cost of money, interest rates; predictions and contingencies for the occurrence of unusual events, as in disaster planning; and predictions of how economic, statutory, and social change may impact on the maintenance operation. The maintenance manager can deal with these uncertainties with a twofold approach:

(1) by collecting and analysing sufficient information to place as much degree of certainty on the variables as possible
(2) by recognising that a level of uncertainty will remain and introducing *flexibility* and *review options* within the plan.

The information collected to enable the maintenance manager to predict with some degree of confidence will be in the form of condition surveys of the property, details of costs and cost trends in building, and information related to the wider policies of the owners and users, for example the corporate plan. Care in such information gathering will enable the manager to reduce the uncertainty and permit effective planning to proceed.

However, the fact that some uncertainties will necessarily remain does not invalidate the planning process. Risk assessment and risk management techniques are commonplace in management generally and construction in particular[2][3]; and these techniques may be adapted for the purposes of maintenance planning.

4.2 Risk management in planning

The concept of risk necessarily takes the manager from dealing with certainties, to probabilities. Any plan that is based on a certainty of outcome will become invalid as soon as there is any deviation from this certainty. Given the wide degree of uncertainty in the overall planning of maintenance coupled with the lack of direct control the maintenance manager has over many of them (such as the incidence of extreme weather, or macroeconomic conditions), it follows that any maintenance plan will need a degree of flexibility in order to remain valid in changing circumstances. This may be achieved by the use of contingencies, which will come into effect if circumstances change from those reasonably anticipated, and by review procedures, whereby the plan contains within it the mechanisms to review progress in the light of changing actual circumstances.

As an example of these mechanisms, a reroofing project may allow a contingency sum for repairs and strengthening to the roof structure. The extent of such work will only become apparent when the existing covering is removed, but the maintenance manager may predict what the likely extent of such work may be, based on careful pre-inspection of whatever symptoms of failure may be evident, and past experience of such projects. The amount of the contingency sum will take into account the risk of problems, and will allow all but the worst case risk of every roof requiring extensive repairs, which may be highly unlikely in view of the evidence available. Whilst the project is running, a mechanism to review the use of this contingency sum would serve to monitor the extent to which the fund is being depleted as work continues, thus enabling the maintenance manager to test his or her predictions against experience, or to amend the contract execution or sum if the amount of work is running higher than that which the contingency sum allows for, before the situation becomes critical.

The above example highlights the operation of what Raftery[2] defines as the cycle of risk management:

- risk identification
- risk analysis
- risk response

to which the maintenance manager could add a fourth, *risk review*, to feed back experience gained from one project into future similar projects.

Risk identification marks out what factors are likely to change from those predicted. For example, the day-to-day maintenance manager may take long-term weather forecasts to determine the probability of extreme weather in the coming winter to plan strategies, priorities, and budget allocations.

Risk analysis assigns probabilities to each of these occurrences, expressed as a number from 0 to 1 where 0 is impossible, 0.5 an even chance, and 1 represents certainty. It will also assess the consequences of the risk in terms of the seriousness, on a scale of 0 as trivial and 1 as very serious; or as a direct fiscal cost in pound sterling. Multiplied together, the product of the probability and consequence give an expected value which can be used as the basis of prioritising action. The operation of this risk assessment mechanism is illustrated for the case of a particular element of a building in Chapter 1, but is more widely applicable to any uncertain outcome and the seriousness of its consequence.

The risk response will be determined by the expected value, and will take the form of either prioritising, or of providing contingencies for the eventuality. In the case of prioritising, high priority work will be undertaken first; the amount of resource dictating how far down the priority list it is possible to work in the actual circumstances (hence the requirement for a degree of flexibility in planning referred to above). The amount of resource allocated to the contingency will depend on the expected value; a higher value warranting allocation of a greater contingency resource, for example a winter sinking fund to provide against predicted damage caused by wind and snow.

This method of risk analysis can be applied *in extremis* to disaster planning[3]; that is, planning for events which, although unlikely and therefore with a low or unquantifiable probability of occurrence, if they occurred could cause disproportionately highly damaging consequences, for example in the case of fire. Indeed the cost of a fire protection system that may never in fact be used in a particular building during its life may be evaluated in this way, and indeed is by insurance valuers when deciding premiums. In this case, the probability is low, but the expected value very high in view of the life-threatening and high-cost consequences, thus warranting contingency planning. Other uses of risk analysis in planning include the important area of health and safety risk analysis within the doctrine of 'best practicable means', referred to in Chapter 3.

4.3 The planning process

The object of planning is to ensure that work considered necessary is carried out with maximum economy, i.e., that the work done satisfies the criteria for effectiveness and efficiency. 'Efficiency' means how well a particular process is carried out. 'Effectiveness' means to what degree the outcomes of the work (however efficient or inefficient) fulfil the goals of the plan. It should be noted that the existence of any kind of plan presupposes a series of goals or policy objectives, or else the plan is pointless and without direction. The success or failure of a plan may only be measured in terms of the degree to which it has achieved the goals set. In this context, policy precedes the plan, and therefore any plan must begin with a statement of the goals or objectives to be reached, before determining how the goals are to be reached. The goals may be very broad at the outset, such as: 'maintain properties to highest standard available within current resources (i.e. by not raising the rents)'. Under these broad goals or mission statements, more detailed objectives can be set as information on the current state of the properties is obtained through surveys, and other 'environment' information such as available budget, become defined. Any uncertain or variable parameters could be profitably identified and assessed at this stage as part of the risk modelling. The plan itself, as a series of activities in a sequence within a given time frame, may then be drawn up in greater detail, including provisions for contingencies or priorities. When the plan is implemented, variations from the expected may be dealt with according to these priority or contingency provisions. On completion, a review of the effectiveness of the whole process in the light of circumstances may be worthwhile as a learning vehicle to input into future plans. This process is illustrated in Fig. 4.1.

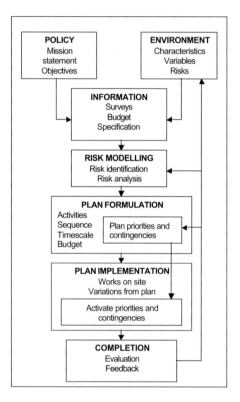

Fig. 4.1   Components of a maintenance plan.

The concept of effectiveness

As stated above, effectiveness may be regarded as the degree to which a plan's objectives are fulfilled. Robertson[4] defines work as 'cost effective' if it is work required as a result of fair wear and tear and is done adequately ('adequately' presumably refers to both quality and cost). Work not required or work done inadequately is regarded as 'ineffective' in that it produces few or no benefits and may attract penalties in the form of additional future costs. Another source describes effectiveness as a measure of actual performance against planned performance. This is too limiting in that effectiveness in the context of the promotion of the overall aims of the organisation is dependent upon the adequacy of the plan. If the plan is in itself defective, then it is difficult to see how fulfilling it can be regarded as effective. The goals set the direction, and the plan itself is a component of how these goals are achieved. The fulfilment of the plan will thus rest in the mechanisms within the plan for achieving an effective outcome. If the mechanisms to deliver the plan's goals are ineffective, the plan will fail.

Sources of ineffective costs

(1) Unnecessary work including:
 - work over and above that required to maintain the building to the specified standards
 - making good the effects of neglect, improper maintenance and misuse of the building
 - rectification of design defects and faulty workmanship in the initial construction.

Harper[5] suggests that rectification is the most profitable point at which to reduce maintenance costs in that it is avoidable.

(2) Uneconomic work resulting from:
 - unproductive time caused by excessive travelling from job to job, waiting for instructions and materials, failure to gain access to premises, inclement weather, etc.
 - improper work methods resulting in more time being spent on the job than necessary and/or waste of materials. The cause may be attributed to imprecise instructions or incompetence on the part of the operative
 - lack of motivation on the part of the operatives
 - inappropriate tendering procedures and contract arrangements in relation to the type of work and prevailing market conditions
 - changes to the nature and scope of the work after commencement
 - lack of an efficient system of recording and controlling costs.

(3) Inadequate work resulting from:
 - failure to identify the true cause of the defect and to specify the correct remedial work
 - improper execution of work as a result of lack of proper supervision, instructions or operative skills
 - lack of adequate safeguards in the contract to ensure that the work is carried out in accordance with instructions and the provision of suitable remedies in the event of failure to comply with the contract.

4.4 Components of a planning system

The essential feature of a planned maintenance system is that failures are anticipated and appropriate procedures devised for their prevention or rectification. It involves having a planned course of action for dealing with the inevitable consequences of deterioration. The plan should be all embracing and lay down measures for dealing with even remote possibilities, e.g. damage caused by severe weather conditions. Whether or not resources are allocated for the rectification of such defects will depend upon the degree of risk involved and the likely effects of delay on user activities.

The question that is sometimes asked as to what are the advantages of a planned approach can best be answered by considering the consequences of an *unplanned* approach. That is, to do nothing until a defect is reported and then request a contractor to do something about it, often with very little control over the work that is done or the price that is charged. This may be described as the reactive or day-to-day approach in that each day presents a fresh set of completely unforeseen problems that have to be dealt with on a purely *ad hoc* basis. It is in fact the antithesis of planning. The following comparison in Fig. 4.2 highlights the relative merits and demerits of the two processes.

Of course, not all jobs can be predicted with sufficient certainty for inclusion in a long-term programme. The object should be to obtain the most economic balance between day-to-day and programmed work. This can be determined by a statistical analysis of the frequency with which particular defects occur and whether or not it would be cheaper to renew all suspect components at the same time rather than carrying out *ad hoc* repairs on a piecemeal basis. Also, planning is not confined to long-term programmes but is equally applicable to the short-term organisation of day-to-day maintenance. The key factor in the management of maintenance is obtaining positive control over the work to be undertaken during any period.

Work is input from two primary sources:

(1) Work initiated by the maintenance department and consisting of larger jobs planned some time in advance. These jobs, while necessary for the long-term preservation of the building, may not have a high degree of urgency.

(2) Work requested by the occupier and consisting for the most part of small jobs which, at least by the occupier, are regarded as urgent.

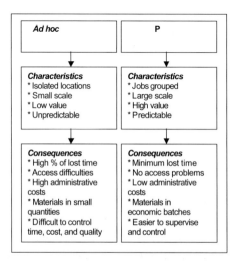

Fig. 4.2 *Ad hoc* and planned maintenance systems compared.

However, if the volume of small jobs requested by the occupiers is too great the frequent withdrawal of resources from programmed work to deal with them will seriously affect the overall efficiency of the organisation.

This particular problem was faced by a local authority where originally all requests for housing repairs came from tenants. The result was that the incoming workload was irregular in quantity, type and location and as a result was difficult to plan and schedule. The effect was that unproductive time was excessively high and often aggravated by the fact that access could not be gained to the premises as lack of planning made it impossible to give advance notice to the tenants. Another danger of relying solely on this method of determining the workload is that tenants will not only vary greatly in their views as to what constitutes a desirable standard, but will tend to report superficial defects rather than more serious but less obvious structural defects.

4.5 Schedule/contingency systems

The maintenance plan must therefore strike an economic and socially acceptable balance between the operation of two complementary and interacting systems – schedule or programmed system and contingency system. These systems and the associated procedures are shown in flow diagram form in Fig. 4.3.

Schedule system

This covers items which tend to deteriorate at a more or less uniform rate and which do not have a high degree of urgency. The procedures take the following forms:

- Scheduling work to be carried out at predetermined times. This includes planned preventive maintenance and applies where the incidence of failure can be predicted with some accuracy or where the periods are fixed by statute or contract, e.g. the terms in a lease requiring painting to be undertaken at fixed intervals.
- Scheduling inspections to be carried out at predetermined times to detect failures or the imminence of failure. In many cases, the exact time of failure is not known with certainty and inspections are necessary to determine whether or not the work is actually necessary. A five-yearly condition survey would fall into this category.
- Scheduling work and inspections to be carried out at predetermined times. This applies where it is possible to predict that certain work will be necessary at a particular time, but an inspection is necessary to determine the extent of any further work. A servicing visit to a passenger lift or to a gas boiler are examples.

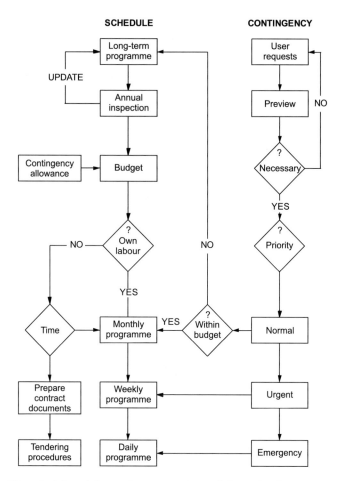

Fig. 4.3   The schedule and the contingency systems of planned maintenance.

Contingency system

This entails a policy of waiting until a complaint is received from the user be-fore taking action. It has been described as casual maintenance but the adjec-tive 'casual' is not appropriate in that it suggests an absence of planning. If the frequencies and types of complaint are analysed, procedures for dealing with the remedial work can be planned even though the timing is uncertain. That is, resources can be parcelled and allocated to deal with the predicted workload.

A necessary feature of this system is the need to introduce a delay period between the receipt of the request and the actual execution of the work. This permits the regulation of the flow of work to the contractor and the grouping of similar items of work in space and time. The main difference between schedule and contingency maintenance is that the lead time, i.e., time between notifica-tion and execution, is greater for schedule maintenance.

Clearly the longer the lead time the more detailed the preparatory work and the less likely are delays as a result of lack of precise instructions or unavailability of materials and plant. The most economic policy is therefore one that maximises the amount of schedule work. However, maintenance is a service and there would be little point in producing a least-cost solution which was unacceptable to the user.

Choice of system

The main factors to consider when deciding whether to treat work under the schedule system or the contingency system are:

(1) The predictability of failure. Components that deteriorate at a known and fairly uniform rate can be scheduled either for inspection or for repair before they actually fail. On the other hand, components that are susceptible to sudden failure, e.g. burst pipes, can only be dealt with under a contingency system. However, incipient faults that may lead to sudden failure may be capable of identification when making the general inspection of the property and work scheduled which will reduce the probability of failure, e.g. lagging exposed pipes.

(2) The reporting delay time. This is the time that is likely to elapse between the time when the defect would just be noticed by a qualified inspector and the time when the occupier would report the defect to the maintenance department. It will depend mainly on the inconvenience that the defect causes to the occupier and is not a measure of the seriousness of the defect. If the reporting delay time is less than the economic period for carrying out inspections the work must of necessity be dealt with on a contingency basis. However, it should be noted that in most cases user requests require previewing by someone who is technically qualified to ascertain the cause of the defect and the scope of the remedial work necessary.

(3) The rate of deterioration of the component and the corresponding increase in the cost of rectification. This must be considered in conjunction with the reporting delay time and whether or not it is the type of failure that is likely to prompt an early response from the user.

(4) The extent to which the user can be relied upon to report significant defects. This will depend upon the nature of the occupancy and the attitude of the user to the condition of the building.

4.6 Factors influencing delay time

As stated earlier, there must be some delay between receipt of a user request and the execution of the work. The factors that determine the permissible delay time include:

(1) *Safety considerations* and, in particular, compliance with statutory requirements. Clearly, failures that constitute a hazard to the occupants or to persons coming on to the premises must be given first priority. The penalties for not doing so may be heavy damages in a civil action or prosecution for failure to comply with statutory provisions. Of course, if the penalties are slight in comparison with the cost of doing the work, then failing to comply could be a calculated risk.

(2) *User satisfaction,* which must be considered even though the defect may be in no way dangerous. It is not uncommon to find a difference of opinion between a tenant and a property owner as to what is an acceptable standard. In some cases, maintenance can contribute only a small part in that the basic cause of the dissatisfaction is attributable to inadequate space or lack of amenities. However, the contingency system does have some psychological advantage in that tenants have a greater sense of participation than would be the case if the whole of the work were planned without their intervention. It can, of course, result in some properties being maintained to a higher standard than others according to the initiative taken by tenants in requesting repairs.

(3) *Effect of failure on the primary activities of the organisation.* This would apply particularly to commercial and industrial buildings that, in this context, are similar to plant and equipment, being distinguished mainly by their very much longer life and the lower probability of complete breakdown.

The buildings are factors of production and any savings resulting from delaying the execution of the work must be balanced against the losses suffered by prolonging interference with the carrying out of basic activities. It is necessary, therefore, to determine whether the defect impinges upon the activities of the organisation, e.g. manufacture, sales, care of patients in hospital, etc., and if so to assess the rate of loss incurred from the state of disrepair.

(4) *Dispersion of job situations.* Travelling time is an important component of ineffective costs and may account for 40% or more of the total cost. Where small jobs are widely dispersed, travelling time can be considerably reduced by grouping jobs requiring the same craft skills according to location. The amount of time saved will depend upon the distance from the depot to the job area and the number of jobs that can be executed in one day (see Fig. 4.4).

For relatively concentrated areas remote from the main depot, such as scattered housing estates, a similar reduction could be obtained by setting up subdepots in convenient positions or by the use of mobile workshops. In both cases the additional cost would have to be justified in terms of the saving on travelling costs and quicker service.

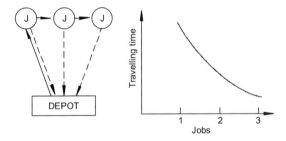

Fig. 4.4 The relationship between job location and amount of work covered.

Even for single buildings, if of a sufficient size, some grouping of jobs may be advantageous in reducing the costs of conveying plant and equipment to the job area and of setting up for the jobs and clearing away on completion.

(5) *Cost growth.* In most cases the longer a defect is left unattended, the more expensive the remedial work, not only to the component initially affected but also to the surrounding parts of the structure. It is not possible to lay down any universal rules and in each case the deterioration characteristics of the affected component must be considered as well as the properties of adjoining components. The cause and effect may be widely separated, e.g. water carried along hidden and devious channels within the structure from one part of the building to another. Deterioration profiles for individual elements would be of only limited value in this respect. Ideally they should indicate the effects of failures on adjoining elements and this will depend upon the unique combination of elements for the building under consideration. Figure 4.5 indicates the general pattern.

(6) *Misuse of property.* There is little doubt that buildings in a dilapidated state are not used with the same care as those that are maintained to good standards. It is recommended in *BRE Digest 132* (1971) that the basic approach should be to maintain to such a standard as to produce a psychological

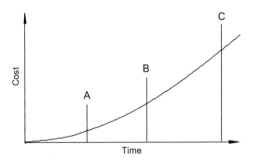

Fig. 4.5 Cost growth related to repair delay. (A) Time when defect just apparent; (B) time when defect likely to be reported; (C) time of repair.

resistance to vandalistic tendencies. Certainly, the maintenance manager should be aware of the temptation provided by a lack of repair and the tendency for faults to escalate as a result of misuse. In addition, good feedback to the designer is essential if mistakes are not to be perpetuated.

4.7 Programming problems

The characteristics of maintenance render the accurate and comprehensive long-term programming of operations impracticable. This is not to say that such programmes do not serve a useful purpose, but that the uncertainties inherent in a large proportion of the work should be recognised and sufficient flexibility built into the programme to permit inevitable modifications. It was remarked in a report by the Tavistock Institute on the construction industry[6] that such programmes can only be based on assumptions about the variety, quantity and timing of the future application of resources and that persistence in unreal assumptions has led to the often uncritical and inappropriate application of techniques of 'scientific management'. In point of fact, the report was referring to new construction, in which the uncertainties are generally very much less than in maintenance. As noted by Skinner[12] such planning systems must be related to the reality of the work actually being done on the buildings and this requires detailed feedback on previous maintenance jobs in order to establish their distribution and frequency. The main factors that militate against accuracy when programming maintenance work are:

(1) The small scale and diversity of a large proportion of the total workload. A study by Reading University[7] of housing maintenance some time ago found that there were about 50 requests per week per 1000 dwellings for small repairs costing on average less than £30. The precise identification of such small jobs in a long-term programme is not possible nor would it be economically worthwhile if it were possible. However, the aggregate amount of these jobs over a period of time shows some constancy and a broad classification according to trade, size and frequency is possible.

(2) The need to determine the best work sequence for a large number of interlocking tasks involving several trades. Thus, where a number of jobs are proceeding simultaneously on different sites the individual job programmes must be timed in relation to one another so that operatives with the requisite skills are available at the right time on each site. The position is aggravated if the work in a particular trade is discontinuous, necessitating two or more visits to the site.

Jobs that are interdependent in that they rely on the same operative carrying out a critical operation at a specific time are particularly susceptible to delays. If operatives are prevented from performing their tasks on one job, they may not be available when required on other jobs and the initial delay, which may have been quite small, escalates throughout the system. One solution is the training of operatives in a number of related trades, so that time spent in waiting on other trades is reduced to a minimum.

(3) The uncertain work content. In many cases, particularly when dealing with old buildings, work is uncovered which could not have been foreseen. This does not necessarily imply incompetence on the part of the inspector, who is usually required to reach a decision on the basis of a visual inspection of the surface effects of the fault. Often it is only when the outer coverings have been removed that the full extent of the damage is revealed. In such cases, changes arising out of a greater knowledge of the defect tend to start a chain of consequential changes, which result in much more extensive work being done than was originally envisaged. Also, this may prompt users to ask for additional work and the initial programme quickly becomes outdated.

(4) The dispersal of sites. This should be carefully considered when programming, from the point of view both of supervision and of the need to redeploy operatives to other jobs in the event of unavoidable hold-ups. Clearly, the geographical grouping of jobs is desirable and also the provision in the same locality of both indoor and outdoor work for operatives of the same trade groups.

(5) Interruptions to the normal progress of the work caused by:
  • withdrawal of operatives to deal with emergencies or to carry out a critical task on some other job
  • inclement weather
  • unavailability of essential materials or plant.

(6) Irregularity of user requests and the extremely short lead time for executing emergency repairs.

In view of the uncertainties, programmes must be formulated at different levels and constantly revised according to the new information that is fed into the system. The levels that can be identified are:

  • Long-term
  • Medium-term
  • Short-term.

4.8 Long-term programmes

  • Quinquennial or longer
  • Annual
  • Monthly, weekly, daily

The object of the long-term programming is not so much to lay down the precise dates when work is to be carried out, as to provide a policy framework. The Woodbine Parish Report on hospital buildings recommends a broad survey of each individual building to establish the major nonrecurring items that are likely to require substantial sums of money within the next five- to seven-year period. Speight[8] describes this as a broad general appraisal in order to formulate policy.

The purpose of the long-term programme is:

(1) To determine the general level of expenditure on maintenance to achieve the desired standards. Inspections will often reveal that there is a backlog of work to bring the buildings up to the specified standard, and this should be recorded separately.
(2) To avoid large fluctuations in annual expenditure by spreading large items and any backlog over a period.
(3) To determine the optimum time for carrying out major repairs and improvements, so as not to interfere with the user of the building.
(4) To determine the structure and staffing of the maintenance organisation and whether it would be advantageous to employ operatives directly to carry out part or the whole of the work.
(5) To gear the maintenance programme to company policy so that it is compatible with decisions relating to the use of the building, e.g. decisions to demolish and rebuild or to move to other premises.
(6) To consider the effect of proposed capital works on the maintenance organisation.

The long-term programme will therefore seek to identify the major items of work over the next five to ten years. The information is obtained from past records showing when major repairs were last undertaken and from inspections of the current physical condition of the various elements.

Painting

Usually the programme pivots around the painting and decorating cycles on account of the predictability of this work and its magnitude – about one-third of the total expenditure on maintenance.

In some cases, the lease lays down specific periods for repainting and these must be complied with, even though they rarely represent the most economic cycles. Where there is sufficient past knowledge of the behaviour of the building and close control is exercised over the quality of paints used and workmanship, it is possible to calculate theoretically optimum cycles. However these periods are essentially averages based on experience and the need for repainting should be confirmed by inspection.

Where appearance is important, it will be necessary to form a subjective assessment as to whether or not the state of the paint work at the end of the economic period will be acceptable. The difference in cost between repainting at the calculated economic periods and at those judged necessary to maintain a satisfactory appearance will represent the value that the occupier attaches to appearance.

In the case of a large building, each elevation should be considered separately in that being exposed to different weather conditions, the paint films would deteriorate at different rates. Against this must be set any savings that might accrue from having the whole of the external painting carried out at the same time rather than piecemeal.

Internal paint films rarely have any protective function and are subject to very little physical degradation over the periods normally considered acceptable. Any attempt to optimise the cycles must therefore be related to the visual impact on the users and the possible effect on their health and working efficiency. Thus, the frequencies adopted for redecorating the internal spaces vary widely according to the type of user and the importance attached to appearance and cleanliness. This is reflected in the following programme, suggested by Luke[9] for the internal redecoration of hospitals.

- Operating theatres: 3 months
- Wards and kitchens: 2 years
- Nurses' home: 3 years
- Offices: 4 years
- Machine areas: 7 years

The work to the various use areas can be phased over a period equivalent to the longest cycle, i.e., seven years in the above example, so that a more or less uniform amount is undertaken each year. In addition, it is advantageous to programme at least part of the internal painting to be executed at the same time as the external painting, to provide alternative work in the event of bad weather.

However, the reasons for adopting different cycles should be carefully examined and justified. In a reported case concerning a local authority the adoption of different cycles for painting and redecorating similar buildings was down to the fact that the standards were laid down independently by the sub-committees responsible for administering the various services. It was suggested that a more rational policy could be achieved if the care of buildings was centralised and placed under the control of a single department. A possible argument in favour of retaining the existing system might be that a department concerned solely with the technical aspects of maintaining buildings might overlook the broader issues. Thus, in the case of this particular local authority, additional money was spent on police houses in order to attract officers from other constabularies to combat the rising rate of crime in the area. The level of expenditure on maintenance was just as much a function of the crime rate and availability of police officers in the area as of the physical condition of the buildings. On the other hand, a central organisation would be in a better position to produce a comprehensive rational policy in line with the needs of the individual users.

Major repairs

These usually involve the replacement or renewal of elements or components for the purpose of eliminating areas of high maintenance costs or restoring lost or diminished amenities. Generally, the elements involved are characterised by a relatively slow rate of deterioration and an increasing need for minor repairs. Roof tiling falls into this category in that, although defective areas can be repaired, there comes a time when the original tiles have deteriorated to such an extent that it is more economic and functionally more satisfactory to renew the whole of the tiling rather than to continue to patch an increasing number of small areas.

However, the rate of deterioration may be accelerated and major repairs precipitated by failure to take early remedial action. Thus, the accidental puncturing of felt roof coverings may start a chain of events calling for progressively more expensive treatments the longer repair work is delayed, e.g.:

(1) patch roof coverings
(2) patch roof coverings and make good internal plaster and decorations
(3) renew damaged timbers and area of felt roofing, make good internal plaster and redecorate room
(4) renew roof complete including ceiling plaster and redecorate room.

The actual timescale of the events is a matter for experienced judgement based on knowledge of the prevailing weather conditions and the quality of the construction. It is remarkable that although treatment at stage 1 is clearly the cheapest, in many cases the condition of the building is allowed to deteriorate to such an extent that stage 4 renewal becomes necessary. Also at this stage there will be the further costs arising from interference with user activities.

It is clear, therefore, that, except where there is a dramatic failure resulting from exceptional weather or other conditions, there is discretion as to the timing of major repairs. However, the deterioration patterns and cost consequences are so ill defined that decisions are largely intuitive, tempered by availability of finance.

Where the property consists of a number of similar self-contained units, e.g. houses, the increased frequency of repairs to a particular element or component might suggest the comprehensive renewal of all the elements or components of that type throughout the estate. There would usually be some element of improvement in this type of work, in that the object would be not only to reduce the incidence of future maintenance, but also to provide a more up-to-date version of the component to give increased user satisfaction. Thus the timing of the work would depend partly on the difference in cost between bulk replacement of the components and piecemeal replacement of individual components over a period of years, and partly on the degree of obsolescence of the components and acceptability to the users.

4.9 Annual programmes

The object of annual programming is to provide a more accurate assessment of the amount of work to be carried out during the forthcoming year and to form a basis for the financial budget. The major considerations would be:

(1) Timing the work in relation to the needs of the organisation so as to avoid interference with the basic user activities and in phase with the overall cash flow pattern.
(2) Providing a uniform and continuous flow of work for all trades in the direct labour force so as to avoid, on the one hand, slack periods because of insufficiency of work and, on the other hand, the need for staff to work overtime at enhanced rates of pay to deal with periods of excessive demand.
(3) Fixing an appropriate timescale for the preparation of contract documents and tendering procedures where work is let to outside contractors and for the advance purchase of materials where the work is to be carried out by direct labour.
(4) Apportioning the amount included in the budget to specific jobs or areas of work for control purposes.

The annual programme would be built up from:

(1) Individual items of painting and repairs brought forward from the long-term plan after a check inspection to ensure that the work is in fact necessary.
(2) Individual items of work disclosed by the annual inspection as being necessary to carry out within the next year. The possibility that such work would be needed should have been anticipated when drawing up the long-term programme and a contingency sum included based on past experience.
(3) Individual items of work proposed by users at the time of carrying out the inspection. Prior to inspecting premises, users should be asked to state their requirements that, if authorised, would be included in the annual programme.
(4) An allowance for work which it is anticipated will be requested by users during the accounting period, but which is not capable of precise definition at the time of the inspection.
(5) An allowance for routine day-to-day maintenance based on past records.

4.10 Short-term programmes

(Monthly, weekly and daily.)

So far, only broad estimates of costs have been produced under the following headings:

(1) Total annual cost subdivided into individual jobs, routine work and emergency work.
(2) Cost of work to be let on contract.
(3) Cost of work to be carried out by direct labour force, subdivided into labour by trades, direct supervision and materials.

The next stage is to allocate the total workload to the months of the year in which the jobs will be carried out. Where the work is to be let to outside contractors, the commencement and completion dates should be entered on a bar chart which should also indicate the dates when the various precontract processes should be initiated and completed (see Table 4.1).

The phasing of contract work should take into account not only the convenience of the user and the ability of the maintenance staff to prepare the necessary tender documents, but also any seasonal variations in prices for the type of work to be undertaken.

Clearly, the day-to-day jobs requested by users cannot be preplanned with the same precision. However, short-term planning is both desirable and feasible. Where these jobs are to be carried out by direct labour a simple method is to use a loading board consisting of a set of pigeonholes extending over five or six weeks with slots indicating the number of each trade at each location. When a work request is received the planner will assess the number of hours for the particular tradesperson and, having checked the availability of the materials, place the order in the relevant pigeonhole for week 1. When all the available hours have been used up the planner will start on the next week, and so on. Some rescheduling is possible according to priorities and to achieve more economic groupings of jobs. The supervisor will be given all the work orders for the forthcom-

Table 4.1 Programme for contract work.

	Apr	May	June	July	Aug	Sept	Oct	Nov	Dec	Jan	Feb	Mar
Job 1				—	—	==	==	==				
Job 2			—	—	—	==	==	==	==	==	==	
Job 3	—	==	==									

— precontract preparation
== contract period

ing week and will arrange the daily programmes according to current labour availability.

4.11 Planned inspections

Inspections are undertaken for a variety of purposes, including:

(1) Preparing a schedule of the facilities to be maintained and their present condition.
(2) Detecting deviations from predetermined standards and incipient faults that may result in such deviations developing before the next inspection.
(3) Ascertaining the cause of deviations, the extent of remedial work necessary to restore to the required standard and prevent a recurrence of the defect, and the relative urgency of the work.
(4) Checking that previous work was done in accordance with the instructions and that the work specified was adequate.

The main advantages resulting from planned inspections are:

(1) An up-to-date appreciation of the overall condition of the property and a corresponding improvement in maintenance records.
(2) More accurate prediction of maintenance requirements and hence better budgetary control.
(3) A greater proportion of the work can be programmed with less reliance on user requests, so permitting the more economic deployment of direct labour or timing of contract work.
(4) A reduction in the risk of breakdowns which may interfere with the use of the building and cause financial loss or inconvenience.
(5) The carrying out of timely repairs will extend the life of certain elements and components and reduce the risk of damage to adjoining parts of the building.

The complexity of buildings and the great variety of possible defects makes it necessary fully to preplan the inspections and to provide comprehensive check lists to ensure that no part of the building is missed. Appropriate criteria should be laid down for each element or component and the type of inspection needed fully defined and documented. This includes details of the form of the inspection, its frequency, the qualifications of the inspector and the feedback procedures. For greater consistency, special proformas or program macros should be prepared for each element and subelement, setting down the essential information required by the inspector with space to record the result of the inspection and the recommended remedial work. In addition, it is advantageous to note the cause of any defect and the urgency with which the work should be carried out.

Location codes should be given to facilitate the later processing of the information. A standardised universally applicable structure for this type of information, Standard Maintenance Descriptions, is detailed in the Appendix.

To facilitate retrieval and computer processing of inspections, the facilities and elements should be grouped according to:

Location

This is of particular importance where the items to be inspected are widely dispersed, in order to devise inspection routes that will minimise travelling time.

The periodicity of the inspections

Building elements and materials generally deteriorate at a slow and fairly uniform rate and inspection at annual intervals is usually satisfactory. If more frequent inspections are necessary, provisions should be made to bring the relevant item forward automatically at the appropriate time. A simple manual 'bring up' card index similar to that suggested for hospital plant could be used for this purpose.

The qualifications of the person who is required to carry out the inspection

(1) *Operatives* where testing and rectification are combined in the same instruction, e.g. checking gutters for accumulation of debris and clearing out if necessary. These jobs involve a straightforward visual inspection that is within the competence of the operative and for which the remedial work can be undertaken immediately without the need for detailed instructions. Control inspections would be carried out at longer intervals of time to ensure that the work was being properly carried out.

(2) *Technicians* with a sound knowledge of building construction and a broad practical experience would be responsible for the general run of inspections. Such persons would normally require special training in the identification of defects, diagnosing the cause and specifying any necessary remedial work.

(3) *Specialists* for inspections requiring the use of instruments or the interpretation of legal requirements or for insurance purposes. This would apply particularly to the inspection and servicing of mechanical and electrical equipment, e.g. lifts. Such persons would be responsible for devising the inspection routines, carrying out control inspections and personally viewing serious defects.

Figure 4.6 illustrates the phasing into the overall work programme of work revealed as being necessary by planned inspections and that requested by users.

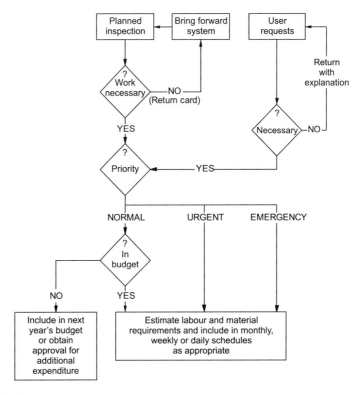

Fig. 4.6   Inspection system.

4.12 Network analysis

The well-known bar or Gantt chart is a simple graphical method of showing the duration of activities. It is easy to understand and is particularly suitable for communicating short-term site programmes to operatives. However, from a management point of view it suffers from the defect of not showing the relationships between different operations. Critical path method (CPM) is better for the larger and more complex jobs in that the network shows the interdependence of the various operations. Thus, in the event of delay, it is possible, from an examination of the network, to determine which operations are critical to prompt completion and to concentrate resources on these operations.

A CPM network represents the sequence of operations or activities in a logical manner. An activity is normally regarded as the work done by a person or group of people but it may also be something that takes only time to perform, such as obtaining approvals. Each activity is represented by means of an arrowed line and starts and finishes at an event (see Fig. 4.7).

The circles represent events and are distinguished from activities in that they are points in time and do not consume resources, whether these be labour, materials or time. The events are numbered and the activities identified by stating

Fig. 4.7   Events and activities.

the start and finish event numbers. The length and directions of the arrowed lines have no significance and serve only to show the logical dependencies of the activities.

Figure 4.7 indicates that the commencement of Activity 2–3 is logically dependent on the completion of Activity 1–2. If it is assumed that these two activities are 'excavate foundation trench' and 'lay concrete foundations' then clearly the concrete cannot be poured until the trench has been excavated. It may sometimes be possible to break down the job into sections which can be excavated and concreted one after the other. The logical sequence of such an overlapping series of activities can be represented by the use of 'dummy activities', as illustrated in Fig. 4.8.

It will be seen that the concreting of the first section, activity 2–3, is dependent on the completion of the excavation of the first section, activity 1–2, and the dummy activity 4–3 indicates that the concreting of the second section, activity 3–6, is dependent on the completion of the excavation of the second section, activity 2–4.

Another way in which dummies can be used to preserve the logic of the network is shown in Fig. 4.9, in which activity C can be commenced only on the completion of A and activity D can be commenced only on the completion of both A and B.

Dummies may also be used to ensure that each activity has a unique reference. Thus if concurrent activities are recorded as in Fig. 4.10 they will have the same start and finish event numbers and be indistinguishable when tabulated.

In order to avoid this, the dummy activity 8–9 is introduced (see Fig. 4.11).

A fault that may appear in a network is that known as 'looping'. Although fairly obvious in the simple example given in Fig. 4.12, it may pass unnoticed in a large and complex network.

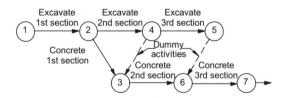

Fig. 4.8   'Dummy' activities.

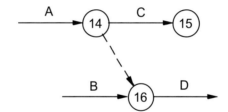

Fig. 4.9   Activity links.

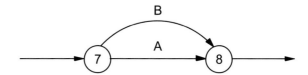

Fig. 4.10   Concurrent activities.

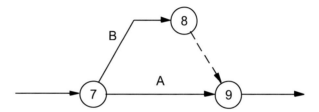

Fig. 4.11   Concurrent activities: use of 'dummy' link.

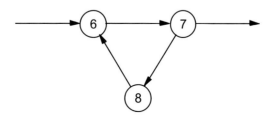

Fig. 4.12   Looping error.

On examining the network one discovers that the commencement of activity 7–8 is dependent on the completion of activity 6–7, the commencement of activity 6–7 is dependent on the completion of activity 8–6, and the commencement of activity 8–6 is dependent on the completion of activity 7–8 – i.e., the commencement of activity 7–8 is dependent on its own completion.

For convenience, it is usual to call the first event No. 1 and then to number the other events sequentially throughout the network. However, any system of numbering may be adopted provided that no two events have the same number.

The first step in the production of a network is to list the activities in roughly the order in which they will be executed. The level of detail will depend upon the purpose for which the network is being produced and the amount of information available at the time. Then for each activity, identify:

- the immediately preceding activity
- the immediately following activity
- concurrent activities.

The network can then be drawn as illustrated in Fig. 4.13.

The next stage in the process is to estimate the time requirements or duration of each activity. Any convenient unit of time can be used such as hour, working day, week, etc., according to the magnitude of the project. The duration should be based on recent experience of similar jobs and allow for normal interruptions to the work. In arriving at the duration it may be helpful to adopt the concept used in PERT (Programme Evaluation Review Technique) in which the most probable time t is obtained from the formula:

$$t = (a + 4m + b)/6$$

where:

m = most likely time
a = shortest anticipated time
b = longest anticipated time

The durations are then entered in brackets under the appropriate arrowed line.

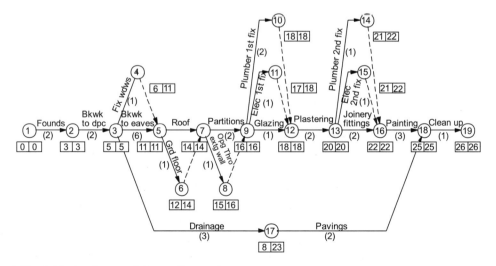

Fig. 4.13   Critical path network for small extension.

The next stage is to calculate for each event the earliest time at which it can be achieved. This is done by making a 'forward pass' through the network from start to finish, adding up the durations of the activities leading up to each event. The earliest starts are recorded in the left-hand box under each event. Thus, assuming that event 1 starts at time zero, the earliest time that event 2 can be achieved is day 3, event 3 on day 5, event 5 on day 11, and so on.

The next computation is to establish the latest time for each event. This is the latest time by which an event must be achieved if there is to be no delay in meeting the final completion date. This is obtained by making a 'backward pass' through the network and is simply a reversal of the method of calculating the earliest event times. Thus, working back from event 19 to event 18, there is only one activity emanating from event 18 and as the duration of this has been estimated as one day it is clear that event 18 must be achieved by day 25 at the latest if the project is to be completed on day 26. Similarly, event 16 must be achieved by day 22, event 17 by day 23, and so on. The latest times are recorded in the right-hand box under the appropriate events.

It will be noted that certain events have the same time for the earliest start and the latest start. These events lie on the 'critical path' which is the longest route through the network. Any delay in completing the activities on this path will extend the overall project time unless compensatory reductions can be achieved in the durations of subsequent activities. Activities not on the critical path will have some spare time or 'float'. Initially activities are planned to start at the earliest time, but to provide a more uniform use of resources throughout the contract period it may be advantageous to delay the start of some of the noncritical activities. This is called 'resource smoothing' and for this purpose, the network is converted into a bar chart as illustrated in Fig. 4.14, so that the total labour requirements for each day or week can be calculated.

Probably the main advantage of critical path methods is that they demand the logical analysis of the proposed work and provide a means of identifying the particular operations on which resources should be concentrated if for any reason the progress of the work is delayed. Although in this case the method has been applied to the site works it could equally well have been applied to the planning and management processes and could provide an effective means of control where deadlines have to be met.

An alternative method of producing a network is the 'activity-on-the-node' system or, as it is sometimes called, 'precedence' diagram. In this system, the activity is represented at the node that is the equivalent of the event in the critical path method. It is generally thought to be easier to understand than the conventional arrow diagram and is frequently used as a preliminary form of analysis. Figure 4.15 illustrates the application of the system to a hypothetical job programme.

The activities are written in abbreviated form in the circles (or rectangles if preferred) and each one is given a unique number. The duration times may be entered subsequently under the activity description.

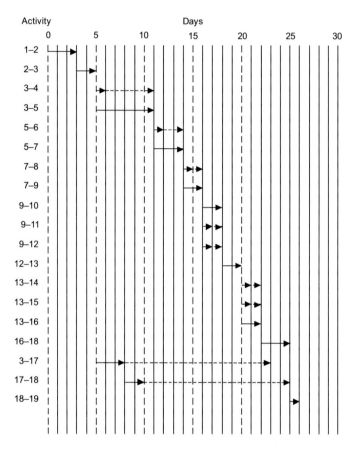

Fig. 4.14   Activity chart showing float time.

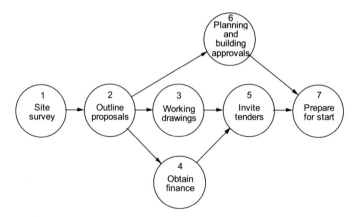

Fig. 4.15   'Activity-on-the-node' system.

4.13 Management by objectives (MBO)

This is a discipline that attempts to define individual responsibilities by setting objectives that have to be achieved within key target areas (KTAs). The object is to integrate an organisation's needs by involving management and subordinates in the process of deciding what to do, how to do it and when to do it. The method provides a rational basis for determining corporate and/or departmental goals and how to achieve them. The basic elements of the method are:

(1) Planning of corporate objectives relating to every area of performance which affects the organisation, i.e., long-term planning.
(2) Planning corporate, departmental and individual objectives which logically derive from the long-term plan, i.e., short-term planning.
(3) Aligning other constraints on managerial action with the set objectives, i.e., budgets, organisation structure, job descriptions, etc.
(4) Co-ordinating and controlling departmental and individual actions so as to contribute to the achievement of corporate goals.
(5) Devising an information system that will indicate the extent to which objectives are being met.
(6) Periodically reviewing goals and objectives in the light of actual achievement.
(7) Involving staff in setting objectives and defining key target areas.

It is important that members of staff concerned should be personally involved in the selection of key target areas and that the targets should be achievable but at the same time present a challenge. Sherwin[11] has suggested that objectives fall into two distinct categories:

(1) Functional performance objectives, i.e., those which maintain predetermined standards and are repetitive such as:
  ● routine maintenance
  ● painting externally every five years
(2) Change objectives, i.e., those that require some distinct cognitive and creative effort and are usually accompanied by some changes in policy, procedures, personnel or investment such as:
  ● reducing backlog of work
  ● increasing speed of response to tenants' requests
  ● improving productivity of labour force by 25%.

Thus the organisation for change objectives is essentially a temporary one, usually requiring co-operation from different groups, but on achievement, the change objectives become functional performance objectives.

It is important to have regular meetings with participants to discuss progress and possible modifications made necessary by unforeseen difficulties. There should be an annual report giving information on such matters as any revisions to the objectives, reasons for failing to meet objectives and details of the steps taken to ensure that the same problems will not occur again. These evaluations are very useful in the long-range planning process in making the department's future objectives more realistic.

4.14 Maintenance audits

Every organisation should have some means of checking its management efficiency. Unfortunately in the case of maintenance, there is no single measure that is appropriate and it is necessary to analyse the answers to a series of questions. Like an accountant's audit the essential feature of a maintenance audit is that it seeks to evaluate the system of internal control, to determine whether or not it is being operated efficiently. Of course, it presupposes that there is a definable maintenance management system to audit.

The audit can be broken down into the following areas:

Technical
- What is the general condition of the building(s)?
- Is the standard of maintenance appropriate?
- Have the defects been correctly diagnosed?
- Is the remedial work satisfactory?
- Are the users satisfied with the quality of the work and the speed of response?

Management
- Is there a maintenance policy?
- What is the organisational structure?
- Are the operating procedures clearly defined?
- What records are kept and are they up to date?
- What percentage of the work is planned?
- What is the percentage of emergencies?
- What is the average backlog?
- Is there an efficient work order system?
- What controls are exercised over time, cost and quality?
- What is the involvement, if any, in the design of new buildings?

Financial
- What is the cost per unit, e.g. per dwelling, per job, etc.?
- Is the cost within authorised expenditure limits?
- What is the return on capital for direct labour?
- What is the ratio between total costs and administrative costs?
- What are the administrative costs per dwelling, per work order, etc.?

Direct labour force

- What is the gross output per operative?
- What is the value of materials used per trade?
- How many jobs are completed within a given period?
- What are the work-hours per job in each trade?
- How are the operatives motivated?
- What is the percentage of unproductive time?

Contractors

- How are contractors selected?
- What is the basis of payments?
- What contractual controls are there over time, cost and quality?
- What is the average difference between final account and tender?
- What is the average delay in completion?
- Is the quality of the work satisfactory?

It will be apparent that operating such a system of control calls for the full co-operation of all members of the maintenance staff, and to achieve this clear guidelines and procedures should be laid down in a maintenance practice manual, as outlined below.

Maintenance practice manual

General

- objectives and policy
- staff duties and responsibilities
- maintenance categories and priorities

Planning and control

- approach to planning maintenance
- work order system
- estimating methods
- performance checks

Inspections

- type of inspection
- period
- check list and report

Budgetary control

- preparation of budget
- cost centres
- variance reports

Execution of work
- direct labour force
  - planning and control system
  - bonus payments
  - stores, plant and transport
- contractors
  - tendering procedures
  - types of contracts and documents
  - supervision and contract administration
  - claims and payment

Information system
- records
- manual inputs versus computer usage
- computer system and software packages
- analysis of information
- feedback and reports

Health and safety measures

Staff recruitment and training

Tenant/user relations

Appendices
- Standard forms
- Flow charts
- References

References

(1) Royal Institution of Chartered Surveyors (1990) *Planned Building Maintenance: A Guidance Note*. RICS Books, London.
(2) Raftery, J. (1994) *Risk Analysis In Project Management*. E. & F.N.Spon, London.
(3) Levitt, A. (1997) *Disaster Planning And Recovery: A Guide For Facilities Professionals*. John Wiley and Sons, Chichester.
(4) Robertson, J.A. (1969) *The planned maintenance of buildings and structures*. The Institution of Civil Engineers, Proceedings Paper 7184S.
(5) Harper, F.C. (1968) Maintenance of Buildings. Paper given at the *Conference on Technology of Maintenance*, Bath, 1968.
(6) The Tavistock Institute (1966) *Interdependence and Uncertainty – Study of the Building Industry*. Tavistock Publications, London.
(7) Reading University (1982) *Housing authorities tendering and contract procedures* (unpublished).

(8) Speight, B.A. (1969) Formulating maintenance policy. Paper given at *Conference on Building Maintenance*, London, 1969.

(9) Luke, R. (1963) Practical aspects of planned building maintenance. Paper given at *International Maintenance Conference*, London, 1963.

(10) Department of Health and Social Security (1968) Planned Preventive Maintenance: A System For Engineering Plant And Services. *Hospital Technical Memorandum No.13.* HMSO, London.

(11) Sherwin, T.S. (1976) Management of objectives. *Harvard Business Review*, May/June.

(12) Skinner, N.P. (1983) *The matching of planning systems to the maintenance work that is actually done.* CIBI W70 Seminar, Edinburgh, March 1983.

Chapter 5
Cost Management

5.1 Maintenance costs

Financial accounting is very important in building maintenance management. Indeed in some organisations, cost is the overriding consideration in assessing the effectiveness of the service. As a general principle, the objective of a maintenance management organisation is to ensure the continuing provision of the required standards and level of service provided by the buildings, at the minimum cost. The cost of building maintenance operations is viewed as an *overhead* on the operations of the building users. An overhead represents a charge on the users of the building, whatever the use may be. In this respect, in business environments there is always pressure to reduce the overhead to the minimum, in order to increase profit on turnover. In residential environments, the overhead is in the form of a direct cost for owner-occupiers and for others as a service charge or rent. The maintenance overhead must be charged against the revenue-generating processes within the building estate, and it comprises two elements:

(1) *Fixed cost overheads.* These include the cost of the provision of the service, such as core staff salaries, accommodation space for the maintenance management operation, plant, vehicles, etc. A characteristic of this type of overhead is that it is incurred largely irrespective of the volume of maintenance work that may be carried out. For example, a repairs helpdesk would incur the same costs whether busy or not. In terms of work carried out, required annual servicing to gas boilers or lifts would also constitute a fixed overhead, being required whether or not the element was faulty or in need of repair.
(2) *Variable cost overheads.* These are costs which depend on the actual amount of work done, and are usually incurred on a unit of production basis. They are directly linked to the actual workload of the maintenance operation. Thus a high volume of repair requests may be followed in the next accounting period by a low volume, with consequent variations in costs.

It follows from the above that the busier the maintenance department is in carrying out the variable cost works, the more efficient it appears to be, as the fixed establishment costs become a smaller percentage of the total costs, and also a smaller percentage of the unit cost of each repair. In this case these fixed establishment costs are analogous to a standing charge for the supply of a utility service such as a phone, where the variable costs represent the actual usage of the service additional to this standing charge. This is known as *overhead absorption,* and the concept also applies to the maintenance overhead as a whole when set against the revenue-generating activities of the user. Jennings[1] expresses this succinctly: 'The process by which cost overheads are attributed to products is termed *overhead absorption*; the alternative label is *overhead recovery*.' The degree of sustainable overhead from the maintenance operation is therefore dependent on the amount of revenue generated by the building users, and is closely linked to the profit, or surplus which may accrue when this and other overheads have been settled. Thus the standard of the maintenance service affordable has a direct relationship to the income of the building users.

Costs may be classified in a variety of ways according to the accounting procedures of the organisation. The basic division is into capital for the provision of new assets and revenue for maintenance and running costs. There are taxation implications in that whereas expenses of a revenue nature can be set against taxable profits, those of a capital nature cannot. Jarman[2] has suggested that, where new capital assets are considered, the financial approval should cover not only the initial cost but also the setting aside of monies to ensure that the necessary funds will be available at specific times for replacement and routine maintenance. However, such reserves are not allowable deductions for tax purposes.

Ray[3] has distinguished between the different types of cost as follows:

(1) *Committed costs,* which represent the after-effect of irreversible decisions taken in the past. Such costs have a mandatory character and include those that are incurred in complying with the terms of a lease or statutory requirements. There is, of course, some element of discretion if the firm is prepared to risk a civil action for damages or a criminal prosecution for disregard of statutory provisions. Other committed costs would be those flowing from specific decisions taken at the design stage, say, to use short-lived finishing materials on the assumption that they will be renewed at predetermined intervals. Again, the costs may be avoided if the occupier is prepared to accept some loss of amenity or a degraded appearance.

(2) *Variable or engineered costs,* which can be directly related to the volume of the primary activity of the organisation, e.g. products manufactured or sales. The difficulty lies in calculating the right amount for a given level of activity. Clearly, where this can be done it provides a simple means of arriving at the permissible level of expenditure on maintenance. However, the rate of deterioration of many building elements is unaffected by user activities and the state of disrepair of these elements would have to be extreme to influence markedly the efficiency with which the activities are carried out.

Where a system of planned maintenance is in operation, the object would be to carry out repairs well before such extreme conditions were reached.

(3) *Managed or discretionary costs,* which require specific decisions in each budget period. This method is criticised on the grounds that there is no scientific method that can determine the right amount that must, therefore, be a matter of judgement. As managed costs represent an 'agreed' amount to spend and not necessarily the 'correct' amount, actual costs below the budget are no reflection of efficiency.

As far as efficiency is concerned, there is clearly a difference between the efficient use of resources in the context of the organisation as a whole and the efficiency of the maintenance department. The minimum amount of maintenance required is that sufficient to support the core production or revenue-generating activities of the building users. If the agreed amount includes unnecessary work, then no matter how efficiently the work is executed, it will represent waste. It is thus a matter of identifying necessary work and this demands the exercise of judgement in assessing the probable rate of deterioration and related cost growth, identifying the cause of the defect and specifying the most effective and economic remedial measures in the light of the building's use. Another alternative is that although the items of work may be necessary, their estimated cost is extravagantly high. Clearly, in such cases actual costs lower than those budgeted would merely reflect poor estimating.

Types of work

Maintenance falls naturally into two main categories.

The first comprises minor items which, although individually of insignificant cost, in total account for a significant proportion of the workload. These are distinguished by the fact that although the timing is uncertain, the total tends to remain constant over a period of time. The inclusion of a lump sum in the budget is open to criticism on the grounds that while this would give financial control, it does not provide a yardstick for performance control. However, the precise identification and costing of each individual item would probably be more costly than any savings it would achieve.

A compromise would be to subdivide the lump sum according to trades, cost range of jobs and frequency. This would necessitate a critical examination of historic costs over a period of years to determine the pattern of expenditure. Checks could then be made at intervals to ensure that the work ordered under this heading was both necessary and efficiently executed. A good example of this in practice is the monthly day-to-day repair budget method used by many social housing agencies. The budget is set slightly higher in winter months to account for greater demand. High priority works always have first call on this budget, and medium and low priority works are undertaken on the basis of what monthly monies remain to be allocated. In this way, the day-to-day maintenance operation may stay within an overall annual budget in the face of variable levels of demand throughout the year. Any other strategy risks either a cumulative shortfall or an overspend during the course of the financial year.

The second comprises major items which, with the exception of jobs necessitated by abnormal conditions, can be predicted well in advance. These items fall into two further categories:

- cyclic work such as external painting and internal decorating, including repairs prior to repainting
- replacement or renewals to eliminate high cost areas.

At the same time, consideration would be given to alterations and additions to improve amenities. Although this work is not strictly maintenance, it is usually the responsibility of the maintenance organisation in that it can be carried out at the same time as associated repairs and by the same operatives. Whether or not renewal is worthwhile in purely economic terms can be easily calculated by comparing the combined annual running costs and amortised initial costs of the replacement with the present running costs. The following information is required for the assessment:

- The present running or repair costs and, where the cost increases with time, the predicted rate of growth based on an analysis of past costs and judgement.
- The costs of replacement including where appropriate an allowance for disruption and loss of service during the period of carrying out the work.
- The anticipated life of the replacement over which the cost can be amortised.
- The running or repair costs of the replacement and, if this is not constant from year to year, the estimated rate of increase.
- Any quantifiable benefits which will be gained from the replacement. Where the benefit is increased and user satisfaction arises, for instance, from the better appearance of the replacement, the benefits can only be evaluated subjectively unless some closely associated cost can be used as an indicator.

The techniques for rationalising repair/replace decisions were described in Chapter 1. When applying these techniques account should be taken of the rate of growth of the repair costs in relation to both the original element and its replacement. Also, there will come a time when further repairs are not possible and the component must of necessity be replaced. The object is to determine the most economic and convenient time to carry out the work.

5.2 Cost indices

Maintenance works by their nature are very variable, to the point where each job may be considered to be unique. This presents considerable problems for the maintenance manager in terms of assessing the cost efficiency of any particular item of work, and for predicting and estimating maintenance budgets. By comparison, costing in new construction is very straightforward, consisting of applying the basic equation:

cost = materials cost + labour cost + overhead absorption.

Although this formula equally applies to maintenance works, obtaining consistent data about unit labour costs in particular is much more difficult. Works may necessitate using or dealing with nonstandard materials or working methods, and other variables such as travelling time, access delays, and clearing-up routines in occupied premises serve to confuse the process further.

Indices are used for the purposes of updating historic cost data, for estimating and adjusting the costs of work carried out at different times to a common level for comparison. Index numbers provide in a single term an indication of the variation against time of a group of related values. The simplest case is where it is required to compare the changes in price level of a single item over a period of time. A specific year is chosen as the base year and the price in that year equated to 100%. Prices in later years are then expressed as percentages of the base year price.

In many cases it is necessary to reflect the general movement of complex variables which consist of groups of items which have different variations. An example of this would be an index to reflect the changes over a period of years of the cost of carrying out maintenance work. One way in which this can be done is to analyse the work over a period of time and identify the major components of cost. This involves determining the proportions of the total cost attributable to labour and each of the principal materials. Then by weighting the separate indices for labour and materials according to their relative importance a combined index can be obtained.

Example

It is required to determine the average increase in the basic cost of maintenance work from 1975 to 1979.

Calculation of combined index for materials

Base year 1975 = 100

Material	Index for 1979	Percentage rise over base year	Weight (percentage of total annual cost of materials)	Product (percentage rise (column 3) × weight (column 4))
Timber	169	69	15	1035
Slates and tiles	182	82	5	410
Cement	197	97	10	970
Plaster	164	64	10	640
Glass	176	76	2	152
Paint	160	60	33	1980
Bricks	180	80	10	800
Copper pipe	194	94	3	282
Sanitary fittings	188	88	3	264
Ironmongery	185	85	2	170
Sand and gravel	214	114	2	228
			95*	6931

Average percentage increase in price of maintenance materials = 73 from 1975 to 1979. Index for 1979 = 173
* Remaining 5% miscellaneous materials.

The total average increase of labour and materials can be found in a similar way.

	Index for 1979	Percentage increase over base year	Weight (found by analysis of year annual costs)	Product
Labour	172	72	65	4680
Materials (weighted)	173	73	35	2555
			100	7234

Average percentage increase in price of maintenance labour and materials from 1975 to 1979 = 72
Index for 1979 = 172

A general criticism of this type of index is that it has to take into account a number of accumulative factors that are difficult to measure and combine. In particular, the productivity of labour that has an effect on the real cost of labour is difficult to assess. Also the pattern of costs is likely to change over a period of time and as this will affect the weighting structure of the index, it will be necessary to start a new series when the change becomes pronounced. This might well be accompanied by the introduction of new materials that did not figure at all in the original list and by a decline in the importance of other materials that previously had a significant effect on total cost. The use of alternative materials in this way might be the result of price increases in traditional materials and would necessitate a change in the weightings of the individual indices to reflect the effect on the combined index of such innovations. It will be appreciated that an index is only an indication of the average movement of prices and cannot reflect the particular difficulties surrounding an individual job.

An alternative method of constructing an index is to base it on the recorded costs of carrying out a standard unit of work at different points in time. This can be applied where the same job has to be carried out at regular intervals under similar conditions and has a constant work content. It is essential that the same variable should be recorded from time to time in order to give a true comparison. For this reason the method is most suited to well-defined single-trade jobs such as painting, where a typical job would be selected and the quantities of the financially significant items measured and priced out at standard rates. At intervals, the items would be repriced at the rates prevailing at the time and the total job cost so obtained compared with the cost derived from the standard rates. The costs at the various points in time would then be related to the standard rate cost, which would be equated to 100. The index would, of course, only be applicable to the particular trade or job under consideration which would be taken as representative of the overall workload of the organisation. The index would reflect not only differences in labour and materials prices but also such factors as efficiency of maintenance planning, materials purchasing and labour productivity.

A simpler method is possible where work is let to a contractor on the basis of a prepriced schedule of rates, in that the percentage additions on the schedule rates quoted by the builder can form the basis of the index. It is, however, difficult to produce a reliable general index for maintenance because of the great variability in the conditions under which the work is executed and the wide range of prices charged by contractors for apparently similar work.

The above examples illustrate cost indices for specific items of material and labour. However the same general principle of cost indices can be applied to many other measures of building cost, for example, average maintenance cost per year per unit of floor area, or average servicing costs for air conditioning plant per year per square metre of floor space served. These indices may be used not only to determine budget provision changes year on year, but also to compare costs for one building or estate with others of a similar type, either within the occupier's organisation or across a property sector. This enables *benchmarking* of costs as a determinant of efficiency of the maintenance operations.

5.3 Benchmarking

Benchmarking is an important technique in cost control, since it offers a method of comparison of costs of similar works or buildings. It consists of comparing similar cost-centres to obtain an average value for a particular provision or operation. However, a precautionary note must be sounded when applying this technique to maintenance works, since because of the uniqueness of maintenance works and of buildings discussed above, straight like-for-like comparisons are notoriously difficult to achieve. Therefore at best, benchmarking must be considered as an approximate method of assessing cost efficiency, and so it is counterproductive to benchmark costs within narrow degrees of accuracy; for example to the penny rather than the pound. The main variables are not only spatial (costs for labour and materials show regional and even local variations), but as illustrated above costs vary through time also. The main parameters for benchmarks in building maintenance can be classed as:

- *Time.* When the cost was incurred. Annual percentage adjustments can be made from a given base year as detailed above.
- *Location.* The economic area, country, region, or even locality may modify average costs. For example, labour is generally cheaper in northern England than the southeast. Even on a local scale, work in central London may attract a cost premium over the same work in the home counties.
- *Estate type.* It is usual to classify cost-centres in building according to the nature of the estate, e.g. commercial offices, industrial, education, rented residential, healthcare premises, etc., even when different estates may occupy similar buildings. This reflects the different user patterns, financial regimes, and management arrangements found in these different sectors.
- *Building type.* Any subdivision here represents broadly different types of buildings, notably the division between high- and low-rise. The differentiation may be taken further, for example buildings with air conditioning and those without, but any further subdivision would tend to progressively invalidate the benchmarking process by hindering the collation of information from a sufficiently large database of similar building types to afford a meaningful benchmark of common costs.

Benchmarked data can be collected from within a single organisation provided that the records go back long enough to give meaningful time-related information, or where the estate is of a sufficient size to furnish a large enough database to average out the idiosyncrasies of particular buildings and uses within it. In this sense it may be used as an internal audit tool to compare the cost performance of different buildings within the estate for the purposes of informing decisions on moves, rationalisations, property acquisitions, etc.

Of far greater use are cost benchmarks provided across a range of different estates; the larger the better, since this enables a finer analysis of individual estate types and building types provided the database is sufficiently large. It also enables organisations to compare their own building running costs with those of competitors or other similar organisations, to assess the level of operational efficiency of their own building stock in an objective manner.

Building Maintenance Information Ltd

For the maintenance manager, by far the most useful of the cost information services on offer to enable effective cost benchmarking is Building Maintenance Information Ltd[4] (BMI; formerly BMCIS), a commercial division of the Royal Institution of Chartered Surveyors. The primary service offered is by subscription from building maintenance managers, who in return for a modest fee and optionally a statistical return on their own operations (never identified individually or by name to preserve commercial confidentiality), receive quarterly detailed breakdowns of maintenance costs, plus special reports on more general comparisons of both costs and practices in building maintenance management, and an annual building maintenance price book. The latter breaks down maintenance operations into what is virtually a schedule of rates format, to enable detailed benchmarking of specific maintenance operations. Costs given are weighted according to time and region to give further accuracy.

More general information is produced periodically on average occupancy costs per square metre per year according to estate type, labour rates and adjustments, materials prices, and special reports on such matters as procurement practices in building maintenance.

Use of these published benchmarks enable the building maintenance manager to assess the efficiency of their own operations against sector averages, and also form an accepted basis for negotiating adjustments on measured term schedule rates or direct labour rates and conditions.

5.4 Financial planning and budgets

The main instrument of financial control is the budget. This is a predetermined statement of management policy during a given period and it provides standards for comparison with results achieved. It has been described as a means of equating available financial resources to planned expenditure. This definition presupposes that maintenance will be planned some time in advance of execution and that resources are limited. Most budgets are established on an annual basis to coincide with the conventional accounting periods, but for maintenance a longer-term budget is desirable, even though some of the items must of necessity be provisional.

The maintenance costs form only a part, perhaps quite a small part, of total operating costs. Indeed, for manufacturing industries the cost of maintaining buildings is usually less than 1% of turnover. In view of the comparative smallness of the expenditure, there is a danger that insufficient attention will be paid to determining the optimum amount to spend on maintenance. This can result in either inadequate funds being set aside for maintenance or an attitude of indifference. The effects of any resulting under-maintenance may not be immediately discernible, but may lead to extensive remedial work being necessary in subsequent budget periods. The long-term plan displays a better picture of the ultimate cost of neglect.

It has been recommended that the budgetary period should be related to the rate of deterioration of a significant element of the building stock. Thus, where circumstances dictate a five-yearly cycle for external painting, the budget period should be based on this. The annual programme may then be seen as a part of a continuing series of work matched to the organisation's cash flow. Shorter-term budgets – say at monthly intervals – are necessary to accommodate unforeseen work or changed conditions.

Budget preparation

The normal procedure is for maintenance departments to produce sectional budgets that are considered in the light of the firm's broad policies, amended if necessary and finally integrated into a comprehensive plan. The financial content of budgets and the close links between budgets and accounts usually result in the controlling and co-ordinating function being exercised by someone with accounting skills. Any pruning to bring the estimates into line with available resources is therefore likely to be to those activities that do not have an obvious and immediate impact on production or which are not shown to produce quantifiable benefits.

In a study carried out by Bath University it was found that it was rare that anything more than a 'guestimate' was made at the budget stage and that where cost estimating was attempted it was elementary. The general conclusion reached was that technical assessments were not sufficiently authoritative to withstand pressures of demands from other users of resources. A similar conclusion is expressed in the Woodbine Parish Report in relation to hospital maintenance organisations. Clearly, if the maintenance budget is presented as a total sum, possibly with large contingency amounts and no indication of the benefits accruing from carrying out the work or of the penalties which will be suffered if the work is deferred, it is possible that funds will be diverted to other apparently more profitable activities.

In such circumstances, much will depend upon the firm's attitude to maintenance and on the personality of the maintenance manager and his or her skill in presenting the case. Often one finds that the maintenance organisation has evolved over a period of years in response to the expansion of the firm's major activities but that the original organisational structure and operating methods remain unchanged in spite of the new demands. The resulting great diversity in the approach to maintenance budgeting with different methods of classifying and recording costs makes interfirm comparisons in this field very difficult. In the majority of cases, maintenance and other occupation costs are fragmented on a functional basis and control is exercised by different people. Thus the interactions between different elements of occupation costs are not made explicit and it is not possible in this situation to devise a coherent maintenance policy.

A lacklustre or too matter-of-fact presentation of the case for an adequate maintenance budget can have a serious impact on the subsequent short- and long-term efficiency of the maintenance operation and on the building users as a whole. As indicated earlier, maintenance is seen as an overhead, and as such, as a drain on profits; therefore the downward pressure on maintenance budgets is considerable. In addition, senior managers may often fail to realise the nature of hidden costs of poor building performance caused by low spending on maintenance, which can be considerable when compared to the maintenance budget itself. For example, the impact of shabby buildings fostering low morale, high staff turnover, and consequent worker inefficiency are difficult if not impossible to track on a balance sheet, though the effects on the overall success of the organisation may be pronounced. This tendency in business to disregard any cost which does not feature on the profit and loss account militates against good maintenance, yet often the maintenance manager when bidding for resources fails to emphasise the importance of such hidden costs on productivity, morale, and even health and safety liabilities.

Subdivision of budget

The benefits of operating a system of budgets are not maximised unless procedures for budgetary control are instituted. The budget should therefore be subdivided in such a way as to permit both financial and performance control. Thus, in addition to stating estimated costs it should also detail the plan of activities and lay down standards against which actual performance can be measured and deviations identified for investigation. It is important that the budget should be properly and clearly presented in order to convince senior management that the proposals have been properly thought out and are both feasible and necessary for the achievement of the overall objectives of the organisation. The maintenance budget should therefore be comprehensive and show clearly the relevance of the proposed expenditure to user needs and the long-term overall policy of the organisation. In particular, as detailed above, it should make explicit the consequences of neglecting to carry out essential or even desirable work.

The following is a typical form of presentation for the maintenance budget.

Title
Maintenance budget (period).

Statement of policy
An outline of maintenance policy and the relevance of this policy to the broader objectives of the organisation. In particular any changes in maintenance policy should be stated, giving reasons and anticipated benefits.

Breakdown of proposed expenditure
The expenditure should be subdivided in various ways to indicate the nature of proposed work and the associated costs.
Possible classifications are:

- by type of costs
- by type of work
- by location
- by method of execution
  - o direct labour – break down into work-hours per trade, materials and plant
  - o contract – indicate types of contract to be adopted.

Discretionary items
This section would cover major works that are not strictly maintenance but which are thought desirable in that they would permit the building to be used more efficiently. Such works would include:

- replacements of major components
- alterations and extensions to existing building

An analysis should be given of the anticipated costs and benefits so that the merits of the proposals can be judged.

Cash flow
A programme should be included indicating the timing of the proposed work and showing the cash flow pattern over the budget period.

Supervisory and clerical staff
Information should be given on any proposed changes in the structure of the maintenance organisation together with any reallocation of staff duties. If necessary this would be accompanied by requests for the upgrading of existing staff or the appointment of new staff.

Direct labour force
Details should be given of the following where appropriate:

- number of staff employed and anticipated wage bill
- reasons for any change in the number as compared with the previous period
- purchase of additional plant with reasons
- introduction of incentive scheme

Appendices
- outline of annual inspection report
- outline of long-term programme to show timing of major works
- summary

Variance reporting

A system of reporting at regular intervals should be instituted so that any difference between actual costs (including committed costs) and those forecast in the budget can be analysed, the reasons ascertained and, if necessary, corrective action taken. Staveley[5] suggests that control should be exercised from a study of monthly statements of expenditure against budget allocation and recommends that:

If overrunning the budget, either
- defer nonprogrammed work for inclusion in the following year's budget, or
- reduce the scope of programmed work.

If under expenditure, bring forward major repairs
Presumably, the nonprogrammed work referred to is that arising from user requests for which a general allowance was made in the budget. However, it would be difficult to apply these remedies during the latter part of the budget period. In particular, bringing forward major repairs might result in inadequate time being available for the proper planning of the work or in the work having to be carried out at an inconvenient time. Before deciding upon the remedy, the reasons for the variance should be carefully analysed. These may be:

(1) Tenders in excess of the estimate as a result of
- inaccurate estimates
- unforeseen increases in wage rates or materials prices
- change in market conditions affecting tendering climate
- delay in starting the work resulting in tenders being invited at an unfavourable time.

(2) Increase in scope of the work as a result of
- variations during execution period or additional work revealed when surface coverings removed
- higher proportion of user requests than anticipated
- exceptional weather conditions or abnormal use.

(3) Inefficient organisation and planning of work resulting in
- higher proportion of unproductive time
- lower output during productive time
- materials wastage.

It will quite frequently be apparent that an increase in the budget allowance would be more appropriate than deferring the work, in that the cost of not doing essential repairs might well be greater than the cost of carrying out the work during the current budget period. Also, monthly fluctuations may be misleading and cumulative trends are likely to be more informative. Maintenance expenditure will not normally be spread evenly over the year since not only do user requirements often dictate the carrying out of major repairs at particular times, but also it may be possible to take advantage of seasonal variations in contractors' price levels. The monthly divisions of the budget should therefore be related to the work programme and not, as is usual accountancy practice, one-twelfth of the annual amount. Any adjustments to the work programme necessitated by unavoidable delays should be reflected in corresponding amendments to the monthly budget. Account should also be taken of the time-lag between execution of the work and payment, and of the fact that some part of the monies due to a contractor is usually retained for an agreed period after completion as a protection against defective work.

5.5 Annual budgets

The chief financial method of regulating maintenance costs is the annual budget. The budget is usually a part of the larger annual budgeting and financial auditing procedures within an organisation. One unwelcome effect of this on the maintenance management operation is that the budget period runs from the start of the tax year, which is in early April. The consequence of this is that the maintenance procurement may be adversely affected by being distorted into this time frame. This manifests itself in a number of ways:

(1) On the day-to-day or contingency budget, the period of highest and most variable demand, the winter months, comes at the end of the financial period. This means that careful husbanding of resources may be required to ensure a sufficient contingency remains to meet eventualities during this period. If this contingency amount is insufficient because of a bad winter, a budget shortfall will occur. If on the other hand a surplus remains after a mild winter, the surplus amount may be deemed to be unnecessary, and deducted from this and from future budgets. The end of the tax year can therefore trigger a spending spree by maintenance managers anxious to use up their budget allocation on lower priority works which may have been postponed to provide the necessary contingency. The difficulty with this is that the works have then to be carried out during February in order to be completed and invoiced in sufficient time to meet the year-end deadline, if no carry-over provision to meet such commitments has been agreed with the finance director. This disadvantages external works in particular, which may be difficult to carry out to an acceptable standard during inclement

weather. Running a monthly budget, in the manner detailed earlier in this chapter, can lessen the impact of this 'year-end spend' syndrome on effective maintenance management.

(2) On the planned maintenance budget, large-scale replacement contracts may be delayed in starting because the start of the financial year inhibits the maintenance manager from entering into contractual arrangements before April or May. This may mean that the good weather months of spring and early summer are lost whilst contracts are signed and starting dates agreed. Work is subsequently pushed back into late autumn and winter. External joinery repairs, heating system replacement contracts, and external painting contracts are particularly vulnerable in this respect.

Frequently, annual budgets are formulated on the basis of the previous year's outcome budget (see above under 'variations'), with an allowance for general price inflation. Although common practice, this is seriously flawed as it looks backward to past performance of the building stock in terms of levels of repairs etc., rather than forward to future need. In a situation of an ageing stock, i.e., one in which newer buildings are being procured or acquired at a rate which means the overall average age of the stock is increasing year on year, this may lead to a shortfall in the budget as the rate of incidence of building element failure increases with age. However, because of the difficulty of accurate prediction of future deterioration, this method is often adopted as the easiest gauge of future need. It may still be used provided that the budget analysis also looks at repair backlog at the end of the year, and at future possible liabilities. As a variation of this error in looking at past cost performance as the sole guide to future provision, a particular problem can become evident in the case of impending failure or obsolescence of a particular element in the forthcoming year. For example, the heating systems of an estate of houses built in the late 1970s will merely require a service each year until a point is reached when either the incidence of breakdowns starts to rise as they reach the end of their operational life, or spare parts begin to become difficult or impossible to obtain. In such a case, a major capital project to replace the ageing systems will become necessary, provision for which may not have been made in a budget derived simply from analysis of previous performance and maintenance needs.

In preparing an annual budget, it will only rarely be possible to obtain finance for all anticipated repair needs because of the downward pressure on maintenance budgets discussed above. Therefore a system of priorities will be required in order to schedule those works which are affordable in any one financial year according to repair need. The priorities should reflect the seriousness of the repair need from the twofold criteria of cost-growth if neglected, and degree of interference with the building use. The actual allocation of a budget will therefore serve to draw a line under those repairs for which the finance is available, and those of lower priority which will have to be deferred. In presenting a budget case to senior management it can therefore be informative to indicate just what a given level of budget provision will achieve, and equally important, what repair needs will have to be postponed. It is also important that those repairs which are not done in any one year have their priorities reassessed prior to the new year budget bid, otherwise if they continue to retain their existing low priority status they may never be done until serious failure occurs as a consequence of repeated postponement.

The consequences of deferral of essential repairs and services as a result of short-term cuts in the annual budget can be serious and long-term. In times of cash flow difficulties, maintenance budgets present a tempting target for economies, since the budget cut can be made with little apparent consequence in the short term, since the progressive deterioration of buildings takes place over many years rather than a single year. Consequently the maintenance manager may find it difficult to argue for a budget retention in times of hardship within the organisation, when cuts in other support services such as catering and company vehicles may be having a much more noticeable and immediate effect. However, such economies can trigger a 'spiral of decline' in the state of the portfolio, as illustrated in Fig. 5.1, where the consequences of deferral lead to a multiplier effect as the lack of repair triggers further deterioration and alienates the building users.

This is exactly the scenario many local authorities found themselves in during the 1980s and 1990s, when draconian cuts in central funding, and disastrous changes in revenue and taxation patterns, starved the ageing housing estates of essential long-term maintenance. This resulted in a premature deterioration to what became, for many estates, an irrecoverable situation. This illustrates how short-term thinking may trigger fundamental long-term problems in building maintenance.

5.6 Long-term costing

Long-term estimates

Long-term estimates may extend over a number of years and be required for a variety of purposes, including financial planning on a national scale for public buildings such as hospitals or schools or long-term maintenance programmes for individual buildings. The characteristic feature is that the precise nature of the individual items of work is not known and the estimate must be based on the average cost of maintenance related to some parameter of the building or buildings in question. The benchmarking data referred to above may be of particular use in making such forecasts. Methods include the following.

Financial criteria

Various ways have been suggested of relating maintenance costs to other costs or receipts. These include expressing maintenance costs as a percentage of:

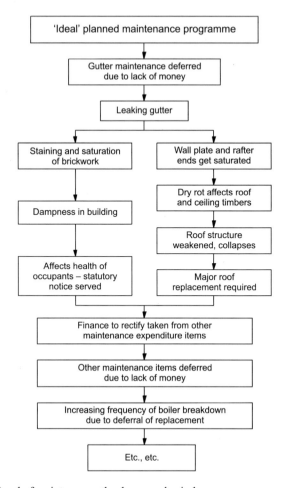

Fig. 5.1   Deferral of maintenance: the downward spiral.

Construction costs

This should be the current replacement value in that the initial cost of construction would give a misleading picture in this context. Unfortunately, replacement value is often merely a 'book' figure and does not accurately represent the actual cost of rebuilding. Also the percentage will vary over a period of time according to the age of the building and the different rates of increase of maintenance costs and new construction costs. There will also be a variation between buildings according to the type of construction, e.g. higher initial costs may have been incurred to avoid maintenance, and theoretically, this should give an inverse relationship.

Production costs

These are determined by factors quite different from those that generate a need for maintenance, e.g. product mix, number of shifts, efficiency of production equipment, etc. Indeed, it is possible that a reduction in production costs by greater mechanisation may increase the maintenance costs.

Occupation costs

Again, there may be an inverse relationship between maintenance and certain other occupation costs: e.g. fabric maintenance may, by improving the insulating properties, reduce heating costs.

Profitability

Maintenance is only one of a great many factors which determine profits; others, such as market conditions, level of prosperity, demand for goods produced or sold, advertising, etc., have a much more direct effect. It should be borne in mind that while a fall-off in sales may be temporary, a reduction in maintenance can have costly long-term consequences as detailed earlier in this chapter.

Relating maintenance costs to the above does not take into account the particularity either of maintenance costs or of the other expenditures. It therefore provides only a crude guide where there is some similarity of use and construction.

Costs per unit of accommodation. The unit would be related to the number of people using the building, e.g. pupils for schools, or to functional units, e.g. general wards in hospitals. It is useful for rapid assessments of the probable maintenance expenditure on a national or regional basis for buildings of the same use category. The disadvantage is that the unit costs are not capable of precise adjustment according to the particular problems encountered in individual buildings.

Costs per unit of floor area (m²). Again, this is a quick and simple method of obtaining a rough estimate of maintenance costs. Different use areas will attract different levels of maintenance expenditure and should be kept separate. The unit costs are obtained from past records but require adjustment for age, constructional form, etc., if applied to other buildings.

Costs per unit of volume (m³). As in the floor area method, the different use areas should be treated separately. However, both methods suffer from the defect that there are other factors that influence the level of expenditure. In a study of hospital buildings it was found that the most significant factors in addition to volume were the numbers of buildings and of occupied beds. A formula for estimating maintenance costs was derived by regression analysis in the following form:

Maintenance costs =
x (volume in m³) + y (no. of buildings) + z (no. of occupied beds)

The coefficients relate to one year only and require updating according to the average maintenance costs per unit volume for all hospitals in the year under consideration.

Costs per building element. The assumption is that an analysis of past maintenance costs according to building elements, i.e., external walls, floors, roof, etc., will reveal a pattern of expenditure which will suggest reasonable amounts to spend on each element. The costs are related either to the total floor area of the building or to the area of the particular element. However, there are clearly no absolute or ideal values, and in many cases the elemental costs will be cyclic. A study of expenditure over a period of years will indicate average costs and also the periods for planned maintenance. In addition, a sharp increase in the repair costs for an element would prompt an investigation to discover the reason – whether or not renewal is desirable or if the element is being subjected to abnormal use.

Costs per functional system. This method attempts to relate maintenance costs more closely to the needs of the user and thus to the benefits which flow from different levels of expenditure. It thus provides a more realistic basis for policy decisions than an elemental breakdown, in that the cost of maintaining each facet of the environment is made explicit.

5.7 Estimating practice

The value of cost planning depends to a large extent upon the accuracy with which future performance can be estimated. However, the degree of accuracy should be consistent with the purpose for which the estimate is to be used and the timescale of the predictions. An excessively detailed approach can sometimes result in unjustifiable costs and may keep staff away from other duties. As stated by a spokesperson for a large organisation, management and supervisory staff are liable to become predominantly involved in this process to the detriment of the essential work of management and supervision of the current programme. The conclusion was that it was better to be less meticulous in estimating and to concentrate the maximum effort on management, supervision and control of expenditure. However, the preparation of estimates will create a need to plan jobs in advance and will thus stimulate more effective management.

The accuracy of the estimate will depend upon the amount of information available on:

- the nature and extent of the work
- the conditions under which the work will be executed
- the mode of execution
- the costs of employing labour
- the prices of materials.

Clearly, the longer the timescale of the estimate, the more difficult it will be to predict the above factors and the more uncertain the estimate will be.

Medium- and short-term estimates

The methods so far discussed are attempts to determine at an early stage an 'ideal' amount based on historic costs and related to standards that were considered acceptable. Later, for programming, scheduling and controlling the execution of the work, more accurate estimates are needed based on the actual work that has to be undertaken. The methods adopted include the following.

Analysis

Small jobs can be broken down directly into their labour and material content on the basis of past records of similar jobs. These are usually single-trade jobs that consist of a straightforward uninterrupted work sequence. Having established the labour hours consistent with the assumed job method and working conditions and calculated the material quantities, these are priced out at the rates which it is predicted will apply at the time of execution. Larger jobs must first be broken down into a series of separate items representing discrete parts of the whole. The items usually include both labour and materials and may refer to a single-trade operation or a self-contained multitrade operation. The items are then analysed to determine the labour and material content and priced as for small jobs. Larger and more complex jobs may require to be broken down into the constituent items. It will be seen that the analyses consist of three main elements – labour, materials and overheads and profit.

Labour
Labour times are obtained from records of past jobs, time studies or activity sampling surveys or, if the information is not available within the organisation, from one of the published price books such as the BMI *Building Maintenance Price Book*[6] or Spon's[7]. Great care should be taken when using published prices in that labour times for maintenance operations vary between very wide limits. The main factors that should be taken into account are:

Travelling time. The cost of a small maintenance job will depend to a large extent upon whether it is a small isolated job remote from the depot or one of a series of similar jobs which can be carried out at the same location.

Access to the building. A good deal of time may be wasted if repeated visits have to be made before gaining access to the building.

Working conditions. Apparently similar jobs can present very different problems to the operative according to the height at or position in which the work has to be carried out.

Degree of repetition. Clearly a familiar frequently occurring job will take less time than one which the operative has not done many times before.

Timing of work. If the work has to be carried out outside normal working hours allowance should be made for overtime rates.

Pre-inspection. In some cases it may be necessary to pay a preliminary visit to the site to ascertain what has to be done.

Preparing and cleaning up on completion. The labour constants given in price books assume that operatives can commence the work immediately and that when they have completed the job they move straight on to the next job. A casual observation of the behaviour of workers will reveal that this does not happen in practice. In an occupied building it may be necessary to remove carpets and furniture before starting, the progress of the work may be affected by the need to keep down dust and noise, and on completion the premises must be left clean and tidy.

Labour on-costs. The all-in hourly rates may be calculated in accordance with the Chartered Institute of Building *Code of Estimating Practice*[8]. The additions to be made to the basic wage rates for building operatives are:

- bonus
- overtime
- cost of living supplement
- sick pay
- tool money
- holiday credits
- national insurance and pensions
- Construction Industry Training Board (CITB) levy
- severance pay
- employer's liability and third party insurance
- trade supervision
- Joint Board supplement.

Currently the above allowances add about 66% to the basic wage rate.

Materials
Allowance should be made for:

- wastage and breakage
- quantity – it should be remembered that materials purchased in small quantities will cost much more than the price given in published price books
- transport of materials to site including unloading and possibly storing prior to use

Overheads and profit

The overhead costs are the costs of administering the organisation and include:

- head office costs
- stores and yard
- supervisory staff
- small plant, hand tools, ladders, etc. (the cost of large items is better included in the rate for the particular item of work for which it is used)
- welfare and safety provisions
- profit would be included where the object of the estimate is to forecast the amount of the lowest tender from an outside contractor. It would not be included where the work is to be carried out by direct labour unless the direct labour organisation is treated as a trading organisation when the profit element would represent what is regarded as an acceptable return on capital.

The disadvantages of analytical estimating for maintenance are:

(1) Lack of standardisation of job descriptions make precise identification difficult.
(2) Job times are rarely available in sufficient detail to permit adjustments to be made for slightly different conditions.
(3) The estimated times are usually averages which are of little value for programming and financial control.
(4) The recorded times may not distinguish between productive and unproductive time, with the result that the reasons for wide differences in the times of apparently similar jobs are not made clear.
(5) The recorded times are not necessarily the optimum times and do not indicate the quality of the management.

Judgement

In many cases historic cost data are so scanty that estimates are based solely on 'experienced' judgement. This applies particularly where the supervisor is responsible for estimating and usually takes the form of a total cost based on the number of operatives required and the time they will take to complete. Although this is quick and, in view of the supervisor's familiarity with the work and ability of the operatives, reasonably accurate, it has certain disadvantages: principally that consistency cannot be proved and insufficiently detailed information is provided for feedback purposes. The main justification is that it avoids duplication of effort in that where there is a separate estimator the supervisor must still analyse the job for labour deployment and requisitioning materials. However, it distracts the supervisor from the main task of supervising the work; also, for financial control an independent estimate is to be preferred.

Slotting

Small nonrepetitive jobs may be dealt with by a technique known as 'slotting' or 'bracketing' – that is, jobs are classified within time brackets by reference to typical common jobs. For a suitable base period the total number of jobs and the average time per job are recorded in each of the following groups:

Jobs taking from

    0.1 to 8.0 hours
    8.1 to 16.0 hours
  16.1 to 32.0 hours
  32.1 to 48.0 hours
  48.1 to 96.0 hours

From the frequency and average time of jobs in each group the total time and labour costs for a period can be calculated. The cost of materials can be expressed as a percentage of the total labour cost. The method is appropriate for estimating the budget allowance for contingency maintenance for which the precise nature of work requested by users cannot be forecast, but which nevertheless may follow a regular statistical pattern.

For very small day-to-day routine jobs, an average cost per week can be derived from past records for the different categories of work.

Approach to estimating

The approach to medium- and short-term estimates will vary according to whether the work is scheduled to be carried out by direct labour or by contract.

Direct labour

In addition to forecasting the organisation's financial commitments, estimating has important functions in relation to the control of directly employed labour:

(1) The total estimated labour hours will indicate whether or not the existing labour force has the capacity to do the work and the nature of changes that should be made to its structure and composition.
(2) The estimated job times will form a basis for programming and scheduling labour resources.

(3) The job times also provide a yardstick against which to measure actual performance and provide a basis for bonus targets.
(4) The aggregate job times permit an assessment to be made of the backlog of work that exists at any time so that the response time can be adjusted if it is unacceptably long or uneconomically short.

The emphasis is therefore on estimating realistic times for the jobs based on the known capabilities of the operatives employed.

Contract

Where work is let to an outside contractor the primary object of the estimate is to predict the amount of the lowest tender or, where it is let on a cost reimbursement basis, the final cost. Unfortunately, information from past contract work is apt to be very much less detailed than that from work executed by direct labour. The main reasons for this are:

The method of presenting information to the contractor for tendering purposes
Often contractors are given only a very general idea as to what is required and must use their own judgement as to the scope of the work. Alternatively, they may be presented with a brief schedule of composite items couched in terms that seek to cover every possible contingency. In both cases the result is a lump sum undivided tender which allows for additional work which the contractors assume will be necessary, but which is not identified. Even where more detailed schedules are produced the items are usually unique to the particular job and, in the absence of further subdivision, cannot be applied to other jobs.

However, whereas whole jobs rarely repeat themselves in precisely the same form, they are made up of operational tasks, i.e., short single-trade work sequences, which do tend to be repetitive. It would therefore be helpful if the estimates were related to standardised tasks for which average times could be built up over a period of time.

The contractual arrangements
Maintenance work is frequently carried out on a cost-plus basis and the final account shows only the total labour hours worked and materials used. For jobs of any size it is impossible to allocate the costs to particular sections of the work.

Job method and sequence of operations
Unless the user of the building demands a particular job method, it is usually at the contractor's discretion and will not be known at the time of preparing the estimate.

Tendering climate

In addition to the basic costs of labour and material, it is important to consider the keenness or otherwise of the contractors to secure the contract. This will depend upon the nature and timing of the work and the degree of risk involved. It will also depend in part on extraneous factors such as the volume of other, possibly more attractive, work available in the locality at the time of tendering. Thus the estimate is not in any sense absolute and any change in the programmed dates may render it inappropriate.

5.8 Cost and quality

Performance measures

In order to determine the effectiveness of planning and improved working methods, it is necessary to have some means of measuring performance so that before and after comparisons can be made. In building maintenance the lack of a standard unit of production and the extremely varied conditions under which the work is carried out preclude precise comparisons. To some extent it is possible to compare the amount of the work done by two employees by an examination of their time sheets, but for larger numbers of employees the relationship between work done and time taken must be converted into figures which are more readily comparable. In addition, the measure should have an interpretative function in that it should indicate the nature of errors in past decisions and suggest appropriate remedial action. However, in order to achieve this several related measures are needed covering the various factors which, in combination, determine the level of performance. The measures should be easy to calculate and use and in most cases consist of simple ratios. Such ratios must be used with care as they reflect relative rather than absolute values and the same ratio may be produced by completely different circumstances. For this reason, the following performance measures have been grouped broadly into those that attempt to measure productivity and those that have an explanatory function.

Measures of productivity

Gross output per operative
Gross output per operative can be calculated as:

$$\frac{\text{Total cost of wages, materials and overheads}}{\text{Number of operatives employed}}$$

This gives a crude index of productivity but suffers from the following defects:

- The total cost will depend upon the proportions of the different trades and the relative costliness of the materials they use. It is not suitable, therefore, for intertrade comparisons, e.g. electricians use more expensive materials than painters do and their output per operative expressed in monetary terms is correspondingly greater.

- By paying overtime and bonuses the total cost is increased without necessarily involving a higher rate of productivity.
- Increases in overhead charges give an apparent increase in productivity. In addition, as methods of calculating overhead charges vary so greatly, the figure is not suitable for interfirm comparisons.
- In some cases the cost of work to be done is decided in advance by estimates which are political targets rather than measures of what needs to be done.

Value of materials used

This may be expressed in terms of the total value of materials over the period under review or the value of materials per operative or gang. Although the method gives some measure of the amount of work done, its usefulness is limited as follows:

- The materials used by different trades vary widely in cost and even within a particular trade there will be different proportions of cheap and expensive materials and of labour and materials costs in different periods.
- The greater the wastage of materials, the higher the apparent output.

Orders executed

This method is only viable where there is a high degree of repetition. It otherwise suffers from the following defects:

- The number of work orders completed in the period reflect only the size and complexity of the jobs dealt with. For the smaller, more repetitive jobs, the orders may be banded according to average times taken in the past, and this provides a rough yardstick against which to measure performance.
- Orders may be given in many different forms, e.g. verbally, by letter, printed form, etc., and standardisation of requests is a necessary prerequisite.

Work-hours per job

This method is similar to the previous one and has the same drawbacks. It may be used for frequently occurring jobs of a similar size and character but the job unit is usually too variable to provide a basis for comparison.

Comparison with estimated times

A ratio may be determined as follows:

$$\frac{\text{Total estimated work-hours on jobs}}{\text{Total actual work-hours worked on same jobs}}$$

- The value of the results obtained will depend upon the accuracy of the estimating although, if consistent, the ratio will at least indicate the movement in efficiency. However, estimating accuracy may vary from trade to trade with the result that the ratio will be affected by the mix of trades in any period.
- It is assumed that the estimate accurately represents the difficulties of the particular job whereas in many cases the estimated time is an average.
- A proportion of the work may not be capable of being pre-estimated and post-estimates prepared for this purpose may be influenced by the time actually taken, or deliberately devised to give a certain rate of bonus.

Standard hours

This is a more sophisticated form of the above method in which the work content is expressed in standard hours, i.e., the amount of work that can be performed in one hour by a worker of representative skill and experience motivated by a suitable incentive and with allowances for relaxation and other contingencies. Ideally, the times should be obtained from work measurement but if this is impracticable, by estimating based on past records or by experience. The measure may be expressed in different ways, e.g.:

$$\text{Performance factor} = \frac{\text{Standard hours of work produced}}{\text{Actual hours expended on the work}}$$

$$\text{Cost per standard hour} = \frac{\text{Total cost of labour employed}}{\text{Standard hours of work produced}}$$

Bampton[9] suggests that the common unit should be the gross cost of one hour's actual work on site. The method involves finding the percentage of unproductive time using activity-sampling techniques and then deducting this from the total hours worked to arrive at the number of productive hours. The total labour cost for the job including bonus and overheads is then divided by the resulting number of productive hours.

$$\frac{\text{Gross cost of 1 hour's}}{\text{actual work}} = \frac{\text{Wages} + \text{overheads} + \text{bonus}}{\text{Actual hours worked} - \text{unproductive time}}$$

Accounting ratios

A rough indication of the variations in maintenance costs may be obtained by expressing the total annual maintenance costs as a percentage of replacement value or some primary function of the firm, e.g. production labour costs, total occupation costs, sales, etc. However, in all these cases there are other factors apart from maintenance efficiency that will affect the percentage. For example, the ratio between maintenance costs and replacement value (assuming a realistic figure is calculated annually) will be affected by the disparate rates of increase of maintenance and new construction costs.

Measures of planning efficiency

These are complementary to the productivity measures, which they seek to explain in terms of the consequences of management decisions. They all presuppose a planned approach to the organisation of maintenance work and indeed are only meaningful in such a context. The measures used for this purpose include:

Degree of scheduling

This may be calculated as follows:

$$\frac{\text{Total direct hours on schedule}}{\text{Total direct hours available}}$$

Clearly the aim of a planned maintenance system must be to predict and schedule as large a proportion of the workload as possible. The many small day-to-day routine tasks may be scheduled by the total times per period rather than individually. Variations in the ratio therefore indicate either a different proportion of small jobs from that anticipated or that larger jobs had not been foreseen.

If the proportion of scheduled work is too low, the individual jobs should be examined to ascertain whether or not they could reasonably have been predicted. The cause may have been failure to detect incipient faults at the time of carrying out inspections, too long an interval between inspections, or extraneous causes such as exceptional weather conditions or abnormal use. Conversely, a high proportion of scheduled work might indicate that preventive maintenance is being carried out at too high a level and that some work is being done prematurely and therefore at needless expense.

Lost time factor

This may be calculated as follows:

$$\frac{\text{Actual hours productively employed}}{\text{Total time worked}}$$

This ratio will have meaning only if the causes of lost time are made explicit and, in particular, if a distinction is made between time lost as a result of inefficiency or as a result of natural causes outside the control of both the operative and management. The avoidable causes (or at least those that should be examined to see if they were in fact avoidable) should be subdivided into:

- waiting for instructions
- waiting for materials
- waiting to gain access to premises
- travelling time

Supervision factor

This may be calculated as follows:

$$\frac{\text{Total supervision costs}}{\text{Total direct labour costs}}$$

Alternatively, this could be extended to cover the total management and supervisory costs. The introduction of a planned system, especially if coupled to an incentive scheme, will inevitably lead to higher management costs, but the increase should be geared to the consequential savings resulting from improved productivity.

Incentive coverage

This may be calculated as follows:

$$\frac{\text{Total work-hours on bonus work}}{\text{Total direct work-hours available}}$$

The object is to give some indication of the extent of the coverage of the incentive scheme. Ideally, where there is an incentive scheme it should be applied throughout to avoid operative dissatisfaction. However, it is in the nature of maintenance that some of the work will be of uncertain scope and not capable of accurate pre-estimation for the purpose of fixing a target. In such cases, post-targeting may be resorted to, but this is unlikely to provide a stimulus to increase productivity. Also, it is a misuse of the word 'target' if the time is not determined until after the work has been executed, when it will tend to reflect the time actually taken rather than the time which should have been taken.

Service efficiency

Maintenance provides a service to the users of the building, and high productivity and planning efficiency are of little avail if the service is unsatisfactory. The following measures reflect the quality of the service provided.

Delay in executing work orders

The users of the building will tend to judge the efficiency of the maintenance organisation by the promptness with which their requests are dealt with. This may be indicated in various ways:

- By the average time taken to respond to authorised requests during the period under consideration.
- By calculating the delay ratio:

$$\frac{\text{Number of jobs one week overdue}}{\text{Number of jobs completed in same week}}$$

- By the backlog of work orders accumulated up to the end of the period in question. A breakdown of the backlog by trades will also indicate the adequacy of the labour force available and assist in planning. A small backlog may indicate that necessary maintenance is not being reported or that maintenance staffing is at too high a level.

In interpreting these ratios the trend is more significant than the weekly figures.

Complaints from users

Such complaints may be made in many different forms and may be frivolous or refer to matters of basic design which are outside the scope of maintenance. It would be difficult to combine them into a single measure of dissatisfaction but they may point to specific defects in the system that require examination. This is examined in more detail in Chapter 9.

Consequential costs of breakdown

This applies to commercial and industrial buildings in which a breakdown may interfere with production processes or necessitate the closing down of part of the building while emergency repairs are carried out. The costs flowing from a breakdown are just as important as the cost of the remedial work when determining the justifiable amount to spend on preventive maintenance.

External agencies

These are the agencies which affect the cost of maintenance and which are outside the control of the maintenance organisation.

Cost of employing labour

$$\frac{\text{Total cost of maintenance labour}}{\text{Total direct work-hours applied}}$$

The 'all-in' cost per hour will reflect changes in the basic wage rate and the various statutory and other levies that the employer is required to pay. The cost of supervision should not be included, as this is within the control of the maintenance organisation and is better considered separately.

Cost of materials

The movement in materials prices could be represented by a single index appropriate to the particular mix of materials normally used by the organisation, i.e., a compound index derived from the price indices of the individual materials weighted according to their proportionate use. However, the maintenance organisation can exercise some control over this factor by buying in a cheaper market or in larger quantities, or by using cheaper substitute materials.

Time lost as a result of unavoidable causes

Here the main factor is bad weather, although on examination of the circumstances it may be found that some of the time lost could have been avoided by providing alternative indoor work. This measure would be a subdivision of the lost time factor mentioned earlier.

The measures considered so far have been concerned with efficiency and with minimising the use of resources. It is equally important to consider the effectiveness of the work in achieving the basic objectives of the organisation; also whether or not the work was necessary or appropriate, and the quality of its execution.

Quality control

In minimising cost, it is important to ensure that the quality of the work done does not suffer. This is a supervisory function and is performed by inspecting work during execution and on completion to check that the materials used and standard of workmanship are appropriate. It should be noted that whereas for contract work there is usually a clause in the contract requiring the contractor to make good defects appearing within a specified period, there is no such remedy for defective work done by directly employed labour. The important factors are as follows.

(1) Creation of conditions which will favour good quality work such as:
   - selecting a reputable contractor of known reliability
   - employing operatives with the requisite skills
   - planning the work so that it is not affected by adverse weather conditions
   - arranging, as far as possible, unimpeded access and suitable working space and conditions.
(2) Clear instructions as to what has to be done and how. Where a specification is provided for contract work, it should describe the various criteria that must be met if the work is to be judged acceptable. Care should be taken to ensure that the specification requirements are in fact practicable.
(3) Inspections during the execution of the work timed according to:
   - the complexity of the work – visits should be timed to coincide with the execution of difficult parts of the work
   - the need to amplify the written instructions – this would apply where the full scope of the work is ascertainable only after preliminary work has been carried out

- the time-span of discretion related to the skill and reliability of the operatives concerned.

The object should be to inspect at critical stages in the progress of the work, to avoid a situation where work has to be redone with the consequent waste of effort and material.

References

(1) Jennings, A. (1995) *Accounting and Finance for Building and Surveying*. Macmillan, Basingstoke.
(2) Jarman, M.V. (1967) Selling maintenance to management – the use of proper costing. Paper given at *Conference on Profitable Building Maintenance*, London 1967.
(3) Ray, H.G. (1969) Budgeting for Maintenance. Paper given at *Conference on Building Maintenance*, London 1969.
(4) *Quarterly Index of Maintenance Costs and other Special Reports*. Building Maintenance Information Ltd London.
(5) Staveley, H.S. (1967) The Planning of Maintenance. Paper given at *Conference on Profitable Building Maintenance*, London 1967.
(6) BMI *Building Maintenance Price Book*. Building Maintenance Information Ltd, Kingston. Published annually.
(7) *Spon's Building Price Index* (published annually) E. & F.N. Spon, London.
(8) Chartered Institute of Building (1983). *Code of Estimating Practice*. 5th edn. Chartered Institute of Building, Ascot.
(9) Bampton, E. (1967) Direct labour or contract. Paper given at *Conference on Profitable Building Maintenance*, London 1967.

Chapter 6

Maintenance Information

6.1 Information system

Information is central to building maintenance management. Indeed, it could be said that the management of information about buildings and their users is the whole of maintenance management, since it would be impossible to procure and direct repairs and servicing without adequate information flows. To achieve these flows it is necessary to set up an information management system, whereby information is collected, processed, and output in a cohesive and effective manner. In building maintenance this is not an easy task, since the sources of information are both diverse and uncoordinated, and the output needs to be targeted towards a wide range of recipients, from board members to contractors to tenants. Kochen[1] has described the functions of an information system as 'planning the behaviour of an organism, of alerting the organism to changes in its environment that signal the need for action, and of controlling action towards the implementation of a plan'. A management information system has been defined by the Institute of Cost and Works Accountants[2] as 'a system in which defined data are collected, processed and communicated to assist those responsible for the use of resources'.

However, the collection of information is not an end in itself and is only of value if the information is applied to control actions towards the achievement of specific aims dictated by policy considerations. It is important, therefore, that the basic aims should be clearly identified and appropriate procedures and techniques developed for the storage, retrieval and processing of relevant information. As stated by Robertson[3] information and information flows are justified by their relevance to decision making and the action which has to be initiated. He points out that the earlier the decision is made the more important and irrevocable are its financial consequences, and yet it is at this stage that there is usually the least information available. Hellyer and Mayer[4] point out another common error in information collection, in relation to condition surveys: 'It is all too easy to collect too much of the wrong kind of information, and to be disappointed with the results obtained.'

The characteristics of building maintenance information can be described as follows.

Diverse

The information arises from a very wide variety of sources: from users, contractors, statutory bodies, accountants, lawyers, architects, direct observation, measurements and gauges, etc.

Unstructured

Unlike other professions such as accountancy or law, there are few if any accepted methods of structuring and organising building maintenance information. Even where practices appear to be common across organisations, close examination reveals that the information structures are similar but not identical or compatible[5]. This requires each maintenance management organisation to construct its own information management system, with the consequence of repeating errors previously made elsewhere, and with the reduced capacity to share development and maintenance costs for the information systems among a number of organisations or in the marketplace.

Low value

The value of a particular piece of information in building maintenance management is often low compared with the time and effort (and hence cost) of obtaining and processing it. For example, the time and effort input into the information management necessary to effect a small but necessary repair such as the repair of a WC ball valve may exceed the cost of the repair itself in terms of time and materials. In the above case, the repair needs to be noted and reported by a user; received at the management organisation; processed and prioritised; input onto the data management system; a works order generated and communicated to the term contractor, who needs to receive the instruction, process it through his own system, instruct the operative, who on completion will need to report the repair complete and prepare an invoice for checking and payment, etc. This low value versus utility of maintenance information points towards the central importance of both efficiency of handling at minimum cost, and of collecting only relevant information, or else the information system quickly becomes swamped with data, making the location and retrieval of necessary information more difficult and costly than it would otherwise be.

Time-dependent

Much building maintenance information is time-dependent inasmuch as it becomes obsolete over time. For example, a condition survey finding will be accurate at the time of inspection, and thereafter will become progressively less accurate as the condition of the building continues to change. An appointment time for a repair may be cancelled if the occupier calls in to say it is no longer possible, and this information must be processed and communicated to the operative before the appointment time or else the information is useless. Essentially, building maintenance occurs in a dynamic four-dimensional environment, which means that the time at which information originated is an important attribute of the information.

Multiform

Building maintenance information arrives in a wide variety of forms: spoken word, text, photographs, plans and diagrams, numeric sums and formula; in both real (i.e. hard copy) form or virtual (soft copy on computer), etc. This requires the information system to be capable of receiving, co-ordinating, storing and outputting this range of information modes within an integrated structure.

Information management in building maintenance has in many organisations evolved from old practices rather than been specifically designed for modern conditions[5], and consequently the systems may not process information with the requisite levels of efficiency. However, as maintenance management is an ongoing process, it can be difficult to re-engineer longstanding systems which at least have the virtue of working reliably, albeit not as efficiently as they may; as the changeover and inevitable bedding down of any new and untried systems may seriously disrupt the operation of the management organisation. This problem of how to update and modernise building maintenance information systems in the light of a rapidly evolving information technology is thus very important, and is dealt with in more detail later in this chapter.

The basic components of a building maintenance information system are illustrated diagrammatically in Fig. 6.1. The three components, although represented linearly, interact in a complex manner, as one output (for example a works order) generates a fresh input (an invoice, or a message from the occupier about access). Nevertheless, the structure gives a basis to examine and design information flows in a logical and consistent manner.

In this examination of maintenance information, it is necessary to first examine the needs of the users of the information. This will be followed by an investigation of the nature of the information they require, the sources from which it may be obtained, and the model for structuring and organising the information in a logical and cohesive format. The discussion will then move on to examine the integration of the information and how it may be processed and presented in a way which will facilitate its use in the decision-making process. The critical role of developing information technologies (IT) will then be dealt with in conclusion.

6.2 Information needs

In order to ensure a consistent approach it is desirable that the basic information collected should be capable of being processed to meet the particular requirements of the main decision makers at different stages in the building maintenance management system. These decision makers are: the financial controllers (senior managers of the occupying or owning organisation), the maintenance

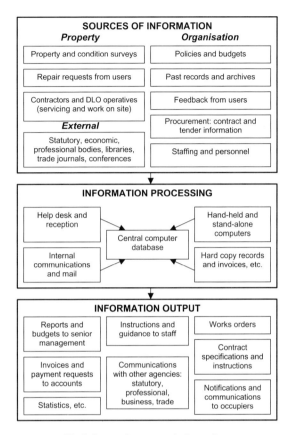

Fig. 6.1   Components of building maintenance information system.

managers at various levels, contractors, operatives, and tenants. Each of these categories of decision makers will require information presented which is:

- concise
- relevant
- easily understood.

The information needs for expenditure and cost control will be examined first, followed by the more general needs of those of the maintenance management system.

Maintenance expenditure control

This is basically a function of upper management and is concerned with the total amount to be spent over a period of time rather than with the detailed costs of individual items, unless these happen to be large and nonrecurring. For this purpose a fairly coarse yardstick would be satisfactory against which to measure the reasonableness of the expenditure proposed or incurred. As stated by the working party on local authority housing,[6] 'We do not think that any properly managed organisation, whether public or private, would incur the management and maintenance expenditure of even a fairly small housing scheme without wanting to know how its expenditure compared with that of other organisations operating in the same field'.

However, it would be unwise to place too much reliance on costs incurred by other organisations in which circumstances might be quite different. The person exercising the function of maintenance cost control is subject to various pressures and his decision to authorise a certain level of expenditure on maintenance will be affected by the strength of the demands from other quarters and, in part, by his own interest in and knowledge of maintenance. It could be a highly individual decision and unlikely to form a reliable basis for other firms in which conditions and financial pressures are different. (See Fig. 6.2.)

Maintenance management

The functions in this area are mainly of a technical nature and concerned with the planning and control of construction resources to ensure that necessary repairs and renewals are carried out with maximum efficiency and economy. The major decisions relate to the following.

Determining standards
For this it is necessary to have information on the overall objectives of the organisation and of statutory and other external requirements so that compatible standards can be fixed. The expression of these standards in qualitative and

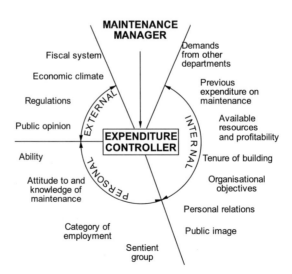

Fig. 6.2   Maintenance expenditure control.

quantitative terms demands knowledge of the effects of varying degrees of disrepair on user activities and levels of visual acceptance.

Planning inspections

Fixing the periodicity of inspections requires knowledge of the rates of deterioration of the building elements so that defects are revealed before they have reached a critical stage. The minimum period will be determined by the inspection costs which should clearly not exceed the cost consequences of failure.

Identifying and specifying the work necessary

This is achieved by comparing the information received on the condition of the building from inspections and other sources with the standards laid down. It demands knowledge of the causes of defects and of the remedial measures which would be appropriate in the circumstances. Often the assessment is based on accumulated experience but there are obvious dangers in applying traditional remedies to new constructional forms and materials. Also, the subjectivity of the assessor in making these determinations requires careful consideration, as shown by Damen[7] and dealt with in greater detail in Chapter 10.

Estimating the cost of the work

As far as possible the estimates should be based on historic cost data obtained from within the organisation for previous similar jobs, but in the absence of such data, costs from external sources and experienced judgement have to be used.

Planning the work

This is mainly in respect of fixing appropriate start and finish times for the individual jobs and requires information on the effect of the timing of the work on user activities, its urgency, the availability of resources and the labour time required for each operation.

Organising the execution of the work

The major decision in this area is whether to employ labour directly for the purpose or to engage an outside contractor: for this, information will be required on the relative merits of these alternatives from the point of view of both cost and convenience. It will include the following.

(1) For contract work, information on contractors' price levels, seasonal variations, and the cost and quality implications of alternative tendering procedures and contractual arrangements.
(2) For directly employed labour, detailed information on wage rates and labour on-costs, output 'norms', job methods, union agreements, prices of materials and plant charges.

Controlling cost, performance and quality

This involves formal systems for the feedback of information on progress so that actual costs and performance can be compared with those predicted and remedial action taken if necessary.

The relationships between the above functions are illustrated in Fig. 6.3.

Design system

The design system embraces all those who contribute to producing a design solution which meets the client's requirements in spatial, environmental and cost terms. In addition to the usually accepted members of the design team, i.e., architect, quantity surveyor, structural engineer and mechanical and electrical engineers, one should include the maintenance manager for advice on the maintainability of the design. Their information needs in relation to maintenance include:

(1) Knowledge of the client's maintenance policy. This relates to general policy in that the details depend on the design finally produced. The probability that repairs will be carried out promptly will influence decisions concerning the required durability of the various elements and components. Where the client owns other buildings an examination of these will give a general indication of his attitude towards maintenance. Also the employment of a direct labour force and the operation of a planned maintenance system would suggest that periodical servicing is likely to be attended to. An important factor which is rarely mentioned in the brief is the planned life of the building and yet this is essential information if the designer is to design for minimum maintenance and running costs commensurate with the initial capital cost.

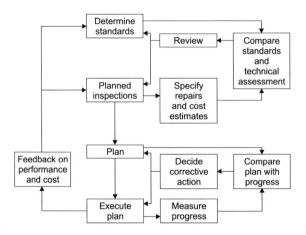

Fig. 6.3   The relationships between maintenance management functions.

(2) Information on the behaviour in use and maintenance characteristics of alternative materials and design solutions. Surprisingly there is very little hard information in this field and of that which does exist only a small proportion is presented in a form which is easily and conveniently usable.

(3) Knowledge of the way in which maintenance and cleaning will be carried out. This information is required so that suitable access can be provided and proper safeguards incorporated to protect the operatives when carrying out the work. It is now a requirement of the Construction (Design and Management) Regulations that a health and safety plan relating to safe access for maintenance is handed over on completion of the project.

(4) Information on maintenance costs for the purpose of calculating the total costs – i.e., initial capital cost plus subsequent maintenance and running costs – of alternative designs. It is doubtful whether actual job costs would be of very much value in this context, since they are affected by many factors other than the intrinsic properties of the design element, e.g. working conditions, tender climate, management skills, etc. It is more important to have some guide as to the frequency with which repairs are necessary and the probable lives under different conditions of the components under consideration.

6.3 Uses of information

It is clear from the foregoing examination of needs that information is used for three main purposes – prediction, comparison, and knowledge (or instruction).

Prediction

This involves the collection of data over a period of time in order to establish trends or the conjunction of information to establish a cause-and-effect relationship, e.g. the bringing together of information on the properties of materials and the environmental conditions to which they will be exposed to forecast their probable behaviour in use.

The nature of the information and the most suitable form of presentation will depend upon:

(1) What has to be forecast, the reasons for the forecast and the amount of detail necessary.

(2) The timescale of the predictions and the degree of accuracy necessary.

(3) The sources of the information, its reliability and the frequency with which new information becomes available and whether the new information supersedes the old or enlarges it.

(4) The importance of external factors and the ways in which these can be anticipated and their effect measured.

(5) The impact of the predictions on other parts of the system and the overall effect of prediction errors.

Comparison

This may take the form of comparisons of internal data for the purpose of measuring performance or of comparing internal data with that obtained from external sources as a check on efficiency. It is important to ensure that the sets of data being compared have been collected and classified on the same basis and that the background conditions are similar.

Knowledge/instruction

In many cases the information is required for the purposes of explaining what has to be done and how it is to be done. The amount of detail and form of presentation will depend upon:

(1) *The purpose for which the information is required.* In some cases the information may be solely for internal use within the firm while in others it may be contained in documents which form part of a binding contract, e.g. drawings, specifications and bills of quantities.
(2) *The knowledge already possessed by the recipient.* Documents rarely contain all the information required by the recipient for an activity and assume a sufficient background knowledge to fill any gaps.
(3) *The method of communication.* Although the primary means adopted are graphical and written, a good deal of information is conveyed verbally and more is being conveyed electronically by email and discs.
(4) *Economy.* The production and transmission of information should be efficient in the use of the resources of the originator and of the recipient.
(5) *Multi-use of information.* In addition to the initial instructional use of contract documents, these also provide a source of information for predictive and comparative purposes.

6.4 Sources of information

There are two points to consider: firstly, whether the amount of information is adequate and, secondly, whether the information that is available is being properly used. Although attention is frequently drawn to the lack of maintenance information there is quite clearly a vast amount in existence which, for a variety of reasons, is inadequately used. Perhaps the main reason is the difficulty in finding relevant information at the time it is required. Faced with the plethora of uncoordinated information, the decision maker tends to use only the most accessible material, combined with intuition. The information may be obtained from internal records or from outside sources.

Internal sources of information

Statements from upper management
It is essential that the maintenance manager should be made aware of the overall policy in so far as this influences the standard of maintenance. This information is required in order to assess the extent to which different standards affect user activities and either hinder or advance the achievement of the organisation's overall objectives. It may be that where maintenance expenditure forms only a small proportion of total expenditure upper management does not think it worthwhile to analyse the problem and is content to leave decisions solely to the technical judgement of the maintenance manager. However, this inevitably leads to an *ad hoc* approach to maintenance and removes the incentive to introduce a planned system designed to secure the most economic use of resources.

Building inventory/project information manual
This database lists the facilities to be maintained, their location and condition, and is an essential prerequisite to planning maintenance. For new buildings a maintenance manual may be provided by the designer, but will require continuous revision in the light of direct experience of the building, and to incorporate any alterations or additions.

A recommended format for this information is illustrated below:

- sources of information:
    - description of building
    - contract consultants
    - contract information
    - authorities
    - subcontractors and suppliers
    - emergency contracts
    - schedule of floor areas and loadings
    - list of maintenance contracts
- general maintenance:
    - regular cleaning
    - general maintenance guide charts
    - general maintenance instructions
    - general maintenance log sheets
    - fittings for replacement
- services maintenance:
    - services maintenance guide chart
    - fittings for replacement
    - log sheets

- o drawing list
- o health and safety plan
- as required under the CDM regulations:
  - o safe access and working platforms
  - o safe methods of working
  - o hazardous materials
  - o other safety requirements

The manual is prepared by the designer largely on the basis of information received from manufacturers and suppliers. If properly prepared, it can be a most useful document but in many cases the maintenance and inspection periods are not related to any defined maintenance policy and are therefore unlikely to be the most economic. Also, in practice, one finds that only a proportion of the items are listed in the manual and that it falls far short of being a comprehensive planned maintenance system.

Inspection reports recording the actual physical condition of the building
The quality of the information will depend upon the skill of the inspector and the degree of forethought given to the setting of test criteria and procedures. The preparation of standard forms or programs to serve as check lists is an important means of guarding against omissions.

User reports relating to inadequacies and faults which detract from the use of the building
Generally information from this source will be incomplete and will require previewing by a member of the maintenance organisation to establish the precise nature and scope of the work involved. Again, it is desirable to standardise the form in which such requests are made and the reporting procedures.

Documents used for the initiation and control of work
The work order is the major source of information on resources used by directly employed labour. Against the standardised description of the job the completed work order should indicate the number of labour hours per craft and the materials used. Suitably coded, the information can be processed to give costs per building, per department, per element, per craft, etc., and if necessary may be subdivided into labour and materials.

Contract documents such as priced bills of quantities, schedules of rates and accounts submitted by contractors on completion of jobs
The amount of information which it is possible to elicit will depend upon the extent to which the work is broken down for the purpose of tender pricing. For new construction, the priced bill of quantities forms a valuable source of information for cost planning, but this is a very detailed and standardised breakdown

into small items of labour and material which are priced individually. For maintenance work there is no similar standard approach to the preparation of tender and contract documents, and in many cases the contractor merely quotes an all-in lump sum or carries out the work on a cost reimbursement basis. In such cases only the total contract sum is recorded and this is not capable of subdivision or adjustment for the purpose of estimating the cost of future jobs.

External sources of information

These are extremely varied and the following account indicates only the more important. Apart from textbooks, the sources include the following.

Government departments

The Department of the Environment, Transport and the Regions (DETR) in conjunction with the Government Statistical Service publishes statistical data of, among other things, the composition of the stock of buildings, the input of the construction industry in terms of type and size of firms, operative trades and earnings and the output by type of client and building. Their website (http://www.detr.gov.uk) is informative and easy to navigate in this respect, though for more detailed information, published reports are still necessary.

Other government departments issue design guides and recommendations on maintenance procedures for the types of building for which they are responsible. The publications of the Department of Health and Social Security are particularly helpful in this respect.

Other sources

(1) The Building Research Establishment has for many years carried out research into and published papers on the technical, economic and organisational aspects of construction and maintenance. Of these publications the *Digests* (first and second series) give useful and concise information on the causes of defects and preventive and remedial measures. Recently they have produced some useful publications on such topics as energy management.

(2) The Chartered Institute of Public Finance and Accountancy is a professional accountancy body for the public sector. The Institute publishes annual statistics of the maintenance costs of local authority housing and also reports on accounting procedures that may be used for this type of work.

(3) The British Standards Institution publishes specifications and codes of practice for materials and workmanship respectively. Unfortunately, few British Standards involve a requirement that there should be experience of satisfactory performance in use and/or the ability to withstand laboratory tests. Also, they lag behind current practice and it takes a considerable time before new materials or methods are incorporated in a BS specification

or code of practice. From the maintenance management viewpoint, useful new Standards include *BS 7543: 1992: Guide to the Durability of Buildings and Buildings Elements, Products, and Components.*

(4) The British Board of Agrément was set up in 1966 to provide technical assessments of new building products. The certificates of approval include an assessment of the probable performance of the product in its intended use and in appropriate cases the maintenance requirements and probable life. They are therefore more informative than BS specifications in regard to the maintenance characteristics of materials but at present cover only a limited range of products.

(5) Manufacturers' literature is produced in very large quantities and varies greatly as to the quality of the information given. Although untrue statements are discouraged by the Trade Descriptions Act 1968, the assessment of claims made for the performance of products often requires skills and equipment not usually available to the designer.

(6) Trade associations have been formed to promote the proper usage of materials by their member firms; most of these associations publish information on maintenance methods and are willing to advise designers and maintenance managers on particular problems.

(7) The National Federation of Housing Associations produces some excellent reviews and guidance notes relating to the maintenance management of social rented housing. Recent examples include *Stock Condition Surveys: a Basic Guide for Housing Associations* and a *Guide for Tenant Participation in Housing Management.*

(8) The Royal Institution of Chartered Surveyors produces a range of articles and publications generated through their Practice Panels. The Maintenance Practice Panel has produced several guidance notes including the *Planned Building Maintenance Guidance Note*, and in conjunction with Owlion Tapes, a number of audio tapes on commercial buildings and schools maintenance. The monthly magazine contains some useful articles and updates, and the bookshop produces an extensive mail-order catalogue of books relating to building construction and property management.

(9) The Institute of Maintenance and Building Management is a specialist professional body well-represented in large property managing organisations, and with an active agenda including a quarterly journal, and a newsletter with book reviews and CPD events.

(10) The Chartered Institute of Building is actively engaged in the furtherance of maintenance studies and for some years has had a special committee to examine the problems in this field. In addition to an extremely useful booklet entitled *Maintenance Management – A Guide to Good Practice*, their maintenance information service has published a number of papers on specific maintenance topics and an up-to-date review of the available literature on the management and organisation of building maintenance.

(11) Subscription services such as the Croner's series titles: *Premises Management, Environmental Management* and *Health and Safety Management,* provide a convenient general reference, updated quarterly. Information is provided both in loose-leaf hard copy, and by CD-ROM disc. These are of particular use as a first-stop enquiry and as a way of ensuring that changes in legislation, standards, etc. are brought to the attention of maintenance managers.

(12) The Building Maintenance Information Service is a subsidiary of the Royal Institute of Chartered Surveyors, and provides a focal point for the collection and dissemination of information on the technical, organisational and financial aspects of maintenance. It offers a means whereby subscribers can exchange their knowledge and experience. One of its primary objects is to encourage better communications between upper management and maintenance management and between property users and the design team. The information is presented under the following headings.

- *General and background information.* This section includes news articles, trends in prices of labour and materials, wage rates and labour on-costs, development of materials and equipment, new legislation and statistical series appertaining to maintenance.
- *Publications digest.* This section gives brief summaries and sources of articles relevant to property occupancy.
- *Case studies.* The reported cases cover aspects of maintenance management and budgeting and control systems.
- *Occupancy cost analyses.* The cost analyses cover the whole range of occupancy costs and are prepared from data supplied by subscribers. They are presented in a standard form.
- *Design/performance data.* The object is to describe the state of elements and components, to examine the causes of failure and to suggest the design correction that might have avoided the failure.
- *Research and development papers.* These are papers on various aspects of property occupancy either by the service or by subscribers.
- *Reference indexes.* The intention is to provide a cumulative reference index of key words to facilitate the retrieval of data.
- *Legislation for property managers.* This is designed to help maintenance managers to keep track of the ever changing pattern of legislation by giving a summary of the statutory provisions which relate to the management of premises.
- *Energy cost analyses.* These give details of the energy consumption of different types of buildings for the purpose of interfirm comparisons.
- *Price book.* This is devoted solely to the prices of maintenance work and should prove extremely useful to all who are involved in planning, budgeting, estimating and checking builders' accounts for this type of work.

The information sheets are sent to subscribers at regular intervals and eventually build up to form a comprehensive library of facts and figures on property occupancy.

Reliability and interpretation of external information

Apart from the fact that many firms do not keep their records in a suitable form or in sufficient detail for the purpose of maintenance management, the information is likely to be uncertain for the following reasons:

(1) There is no universally accepted definition of maintenance and the boundary between maintenance and improvement is often difficult to draw with any precision.
(2) There is a lack of a universally accepted framework covering all divisions of maintenance with the result that maintenance costs from different sources are aggregated under different headings and cannot be re-allocated without reference back to the source documents.
(3) The heterogeneous nature of the work and the smallness of the individual jobs inevitably results in recording errors. Indeed, it is estimated that up to one-third of the annual output on maintenance is not recorded at all and of that which is recorded, the picture is distorted by such factors as different accounting systems or taxation considerations.
(4) Most maintenance is carried out by small firms and as they are the least able to make use of statistical information, they have the least reason to take care in its initial collection.

However, the main problem lies in interpreting the data. The principal uncertainties are:

(1) The data show only the amount spent and, as a rule, do not indicate:
    ● the standard achieved
    ● whether the work done was all that was technically necessary or whether part had been deferred for policy reasons
    ● the quality of the work done
    ● the efficiency with which the work was done.
    Clearly, low costs do not necessarily reflect efficiency; they could equally be the outcome of neglect. Indeed, abnormally low costs are more likely to stem from a policy decision to cut maintenance to the minimum rather than indicating that the building is in first-class condition.
(2) The reasons for expenditure are not usually given and may be completely obscured by the aggregation of costs under inappropriate headings. For both the maintenance manager and the designer it is important to know whether the work was rendered necessary by faulty design or abnormal use or was attributable to normal wear and tear. Without this knowledge it is

not possible to isolate exceptional costs or to understand why they were exceptional.

(3) The costs of deferring work are not ascertainable from statistics of historic costs except in so far as high current costs may be traced to past neglect. In the context of planned maintenance it is essential to consider the cost pattern over a period of years and in particular the effect of alternative periods for cyclic work.

6.5 Information structure

It has been shown earlier that maintenance information is very diverse, and does not at source conform to any form or structure. Therefore it is necessary for the maintenance manager to impose a suitable structure on the information passing through the maintenance organisation. In practice studies have shown[5] that the specific format for any single organisation is likely to have evolved from the historic structure of information management within the organisation as a whole. For example, many maintenance management systems in use with local authority housing departments have as their basis the housing revenue system of asset tracking, to which maintenance data has been appended. This state of affairs is rarely completely satisfactory from the maintenance manager's point of view, since these accreted or amended systems are unlikely to be capable of processing the maintenance information in a comprehensive or efficient way. For example, tendering documentation for schedules of rates may lie outside the scope of the main system and require their own independent information management systems. However, such comprehensive departmental systems do have the advantage of data uniformity, permitting the efficient interchange of information between the maintenance managers and the client department.

Before examining the specific characteristics of a maintenance management system it is first necessary to examine the general principles of information classification.

Classification principles

Classification has been described[9] as the actual or ideal arrangement of those objects which are like and the separation of those which are unlike. The categories should be meaningful to the user and sufficient in number:

- to allow relationships to be built up between relevant groups of information, e.g. elements/standards/costs; maintenance task/resources/cost
- to identify resources with regard to performance, availability and price
- to facilitate the preparation of relevant documents, e.g. inspection schedules/ work programmes/work instructions.

There are two basic types of information to consider:

(1) *General.* Information which is freely available in published form and is useful for a variety of purposes.
(2) *Specific.* Information which has a restricted application and which may be derived entirely from internal sources or consist of a unique collection of general information. It may relate to:
   - the organisation owning or occupying the building, e.g. overall policy, management structure, financial situation, etc.
   - the physical characteristics of the building(s), e.g. constructional form, floor area, etc.
   - the individual items of maintenance work, e.g. material and labour resources, costs, etc.

The primary classification need for general information is to permit ready retrieval of the document, whereas for specific information the ordering should allow manipulation and processing for a defined purpose. However, it is possible for the same information to exist at both levels, e.g. a published standard schedule of work descriptions would be of general applicability whereas a unique permutation of selected work descriptions would be specific to a particular job. Ideally, therefore, both types of information should be contained within the same frame of reference. In view of the widespread use of existing classification systems for general information, it is probably more realistic to think in terms of a complementary but compatible set of categories for maintenance. Figure 6.4 indicates the overall pattern of information flow.

General documentation

This can be classified by any of the conventional library systems. The Universal Decimal Classification (UDC) is the most commonly used and is hierarchical in character. The principle employed is that the total body of knowledge is divided into ten primary groups, each primary group is subdivided into ten secondary groups and these are further subdivided into tertiary groups and so on (Fig. 6.5). The classification thus assumes a tree-like structure with each branch forking into a further series of branches as far as necessary.

The main disadvantage of this system is that the division into classes is fixed and does not permit regrouping under different heads. Thus there is a tendency for related material to be scattered throughout the classes and this creates difficulties when attempting to sort and compare data. Also the system must be conceived as a whole initially and it is difficult to anticipate how much space will be required for future developments.

Faceted classification

A faceted type of classification is usually more convenient for processing pur-

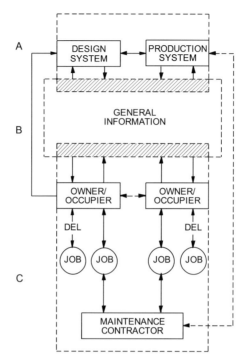

Fig. 6.4 Information flow: (A) specific information relating to new construction system; (B) general body of information; (C) specific information relating to maintenance system. Shaded area indicates overlap of general and specific information.

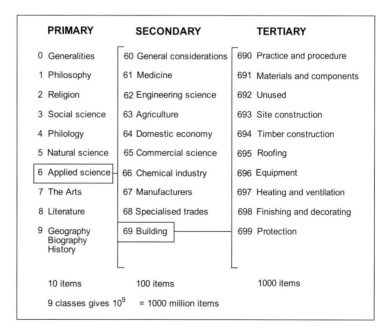

Fig. 6.5 Universal Decimal Classification.

poses. This creates a series of general concepts which, unlike the facets of a hierarchical classification, have equal weight. The advantages of the faceted system are that all possible combinations of characteristics can be catered for and that new concepts and characteristics can be added at any time.

Facet 1	Facet 2	Facet 3
Product	Colour	Finish
Window	White	Natural
Door	Green	Primed
Wall panel	Red	Decorative laminate
Etc.	Etc.	Etc.

The concepts 1, 2, 3 ... are mutually exclusive and may be as numerous as necessary. Similarly the subdivision of each concept may be continued as far as necessary.

The Sf B (Samarbetakommiten for Bygnadsfragor) is a specialised faceted classification which is widely used for filing technical literature relating to construction industry products. The system originated in Sweden, but is now under international control through the CIB (International Council for Building Documentation) which grants licences for publication through national agencies, e.g. the RIBA in the UK. The system was revised in 1968 and renamed CI/SfB. It includes four main tables:

The CI/SfB table system

Table 0	Built environment	a number, usually 2 or 3 digits.
Table 1	Elements	a bracketed number, usually 2 or 3 digits.
Table 2/3	Construction form and materials	a capital letter followed by a lower case letter and usually a number.
Table 4	Activities and requirements	a bracketed capital letter sometimes followed by a number and a lower case letter.

The references are entered in a box, as shown in Fig. 6.6. The detailed tables are given in the *Construction Indexing Manual* published by the RIBA, which also contains information on the method of choosing the correct symbols.

The tables relate mainly to new construction and the space allocated to maintenance is minimal. However, a good deal of the basic technical information is common to both aspects of construction activity and much of the literature from manufacturers and other sources is pre-indexed in accordance with the Sf

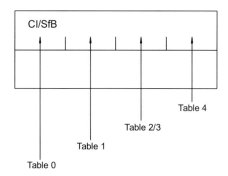

Fig. 6.6   CI/Sf B classification: arrangement of reference tables.

B classification. There would appear, therefore, to be some practical advantage in adopting Sf B at the general level and extending it by the addition of specific maintenance facets.

It is clear that a fair amount of the information is capable of being presented in a standardised form and that transference of this information from the specific to the general level would facilitate communications not only within the maintenance system but also between the maintenance and the new construction systems.

Specific information

This is distinguished from general information not so much by its form as by its particularity. The main purposes for which it is used are the identification, timing, costing and execution of maintenance jobs. These activities both demand and generate information, some of which may be transferred eventually to the general system. The main groupings of this information are shown in Table 6.1.

Table 6.1   General and specific information categories.

| | *Specific* | | | |
General	Organisation	Building(s)	Facility	Operation
Technical	Objectives	Physical parameters	Performance standards	Description and method
Economic	Maintenance policy	Exposure conditions	Inspection criteria	Resource requirements
Management	Financial constraints	User activities	Inspection cycles	Timing
Legislation				Cost

6.6 Standard maintenance descriptions

In new construction, the common arrangement and the standard method of measurement (SMM) provide a common basis for organising information related to construction and interchanging it between different users. The main benefits are:

- few data translation problems between users of the same information, e.g. designers and contractors
- a common system to benchmark costs incurred in different areas, across the industry as a whole
- shared development costs for computer software, by collaboration, or by the creation of a large potential market
- standardised methodologies and routines for costing, building up, etc.

In building maintenance there is no such commonality. Studies[5] have indicated that this is because maintenance management organisations tend to be 'stand alone' organisations which base their information management practices on their historically evolved methods of work, and on the management systems of the client organisation. There is little interchange of meaningful information between different maintenance management organisations. The result of this is the development of maintenance management information systems that are similar, but not compatible. This problem has cascaded into the development of automated systems, as indicated in a recent BMI report[10]: 'of the automated systems used in the UK, there is no clear market leader, and many organisations have developed bespoke computerised systems from their traditional manual systems.' The impact of this diversity has been to limit the effectiveness of building maintenance management practice to respond to the challenges of the information technology revolution of recent years, by restricting the development of automated systems to particular individual users, or small groups of users, with the obvious ceiling this puts on development costs in relation to payback. It also limits the maintenance manager's ability to benchmark maintenance costs against agreed norms because of the difficulties of co-ordinating and averaging differently-scheduled information from a variety of users, as indicated above.

To address this problem, in 1992 the Royal Institution of Chartered Surveyors set in motion research to develop the outlines of a system of information suitable for most maintenance management organisations, as an attempt to parallel the success of the SMM in new construction. The research first examined existing information structures across a range of maintenance management organisations to establish the degrees of similarity and divergence in their systems. Because of the variety of estates and management functions in maintenance management practice, it was found that the standard maintenance descriptions format (SMD) needed to be capable of being tailored to the specific needs of a particular user within the general framework. In particular, the need to keep to a minimum the data needed to run a particular maintenance management operation, as stressed by Spedding[11], required the SMD system to be capable of functioning at any level from the very broad (e.g. the classification of a whole building according to its condition, as part of a broad-brush survey of an estate of buildings, to ascertain priorities), to the very detailed (e.g. a works order to repair a window lock). A hierarchical system of organisation was chosen whereby the data are structured in a hierarchy as illustrated in Fig. 6.7.

Data are entered at the requisite level of the hierarchy according to the particular need. The classification of elements, for example, contains four defined levels of hierarchy, related to built form: for example, building fabric (1); floors and ceilings (2); floor (3); covering (4). The user can specify as a fifth level the particular material or manufacturer (in the foregoing example, of the floor covering), if needed. An outline of the level 1 element hierarchy is given in Fig. 6.8, and the full system is detailed in the Appendix.

The SMD system also seeks to define and co-ordinate the different types of maintenance data used. There are seven classes of information:

(1) the asset register
(2) date and periodicity
(3) building element description
(4) property condition assessment
(5) policy and planning
(6) maintenance procurement
(7) monitoring and record keeping.

Within these, standard 'menus' for information such as location recording, condition ratings, works instructions, and criticality are defined. The benefits of standard 'menus' for such information are as follows.

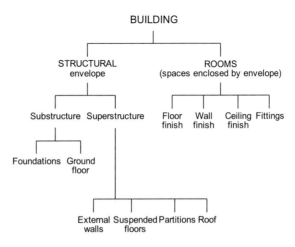

Fig. 6.7   Hierarchical data structure.

```
┌─────────────────────────────────────────────────┐
│          STANDARD MAINTENANCE DESCRIPTIONS        │
│           OUTLINE ELEMENT CLASSIFICATION          │
│  ┌───────────────────────────────────────────┐   │
│  │ B:     BUILDING FABRIC                     │   │
│  │ BN     Foundations and structure           │   │
│  │ BC     Walling and cladding                │   │
│  │ BW     Windows, doors and lights           │   │
│  │ BR     Roofs and balconies                 │   │
│  │ BF     Floors and ceilings                 │   │
│  │ BS     Stairs and ramps                    │   │
│  │ BD     Chimneys, shafts and ducts          │   │
│  │ BX     Sundry fixtures                     │   │
│  └───────────────────────────────────────────┘   │
│  ┌───────────────────────────────────────────┐   │
│  │ S:     SERVICE INSTALLATIONS               │   │
│  │ SE     Electrical systems                  │   │
│  │ SG     Gas and other fuel services         │   │
│  │ SW     Hot and cold water services         │   │
│  │ SV     Mechanical ventilation and air conditioning │
│  │ SS     Sanitary installations              │   │
│  │ SD     Drainage and refuse disposal        │   │
│  │ ST     Lifts and transportation            │   │
│  │ SC     Computer, communications and alarm  │   │
│  └───────────────────────────────────────────┘   │
│  ┌───────────────────────────────────────────┐   │
│  │ A:     ANCILLARY FEATURES                  │   │
│  │ AA     Roads, paths and access             │   │
│  │ AF     Fencing, plantings and external     │   │
│  │ AS     Sports facilities                   │   │
│  │ AX     Other ancillary services            │   │
│  └───────────────────────────────────────────┘   │
└─────────────────────────────────────────────────┘
```

Fig. 6.8 Building and services element classification.

- It permits software developed by one organisation to be directly used by others, by the use of standard menus, which enable automated data handling and processing based on these menus.
- It facilitates common education of maintenance surveyors and maintenance managers, by having an agreed set of criteria. Therefore, the problems associated with surveyor subjectivity (see Chapter 10) are reduced.

The main problems with SMD lie in lack of awareness of the system because there are very limited discussion forums available to maintenance managers, and because of the cost of moving from an existing system to a new one. However, the continued fragmentation of data handling and storage methodologies in building maintenance management, and the consequent lack of sophistication of it technologies currently in use, undoubtedly contribute to an unwelcome 'low-tech' image of the profession.

6.7 Information technology

The continuing revolution in IT development has profoundly affected maintenance management practice, though not as much as it should, as discussed above. The potential benefits of more efficient information handling are enormous in view of the large amount and low unit value of items of building maintenance information, but they have been slow to be realised in management practice. The reasons for this have been much discussed, and include:

The ongoing nature of building maintenance management

Information systems are in constant use, and must often refer to historic data such as tenancy details, previous work done, etc. Any change towards a new system is likely to be accompanied by a dislocation in the delivery of the service; and at its most serious, could result in a systems collapse. Maintenance managers are therefore understandably cautious about bringing in new systems.

Past experience

Past experience of computerisation has shown that it is often necessary to retrain managers and operatives in how to use the new system, to the detriment of the existing service delivery. In some cases, unsympathetic systems have been designed by experts in computing but not in building maintenance practice, which distort the practice by obliging users to conform with the demands the software itself is making in order to function correctly. As Wordsworth and Boughey have pointed out [12]: 'Such systems are characterised by screen menus which require entries to be made in all data fields in order to proceed to the next operation, or which produce reams of printout of information of seemingly marginal relevance.'

Failure of vision

As the BMI Report[10] highlighted, the introduction of IT systems have often merely automated existing manual systems of data collection rather than prompted a rethink of the systems themselves. New IT systems are capable of processing vastly greater quantities of data, provided that the data are collected and assembled in an appropriate format, yet how this capability may be harnessed to provide new information flows in areas previously too awkward or expensive to deal with (such as images and photographic records, which can now be easily copied and appended as files to other structured data) is often ignored.

Cost and risk

The costs and risks of introducing radically new IT in both capital costs, training costs, and dislocation of service are perceived as too high in relation to the short-term benefit to be gained, especially where this may force a radical rethink as to the nature of the management organisation as a consequence. New bespoke systems are very expensive to develop, and are unlikely to work first time; and the lack of a marketplace for common, tried and tested commercial systems available means that changing IT provisions can be a high-risk activity for maintenance managers. Allen and Hinks[13] make the point that, 'within maintenance management departments, the strategic desire to improve customer care may not correspond with the maintenance technician logging requests on to a new computerised system for which he has received little or no training.'

However, despite the above, there are many new IT systems being introduced and developed in building maintenance management. These are only rarely new bespoke systems, but in accordance with the economics of investment in software development referred to above, the BMI Report[10] indicates that: 'the systems available … appear to have developed from three main directions:

(1) *Direct labour maintenance departments.* These systems are characterised by their emphasis on works orders and the accounting required for such items.
(2) *Computer aided design (CAD).* These systems are graphics based and will emphasise space and asset management …
(3) *Engineering plant facilities management systems.* These are based on items of plant being an asset and would have originally been geared towards process plant that is maintained by direct labour.'

The problems with these systems may arise from their structure and operation, initially designed for an activity other than building maintenance management, which may cause the management operation to be distorted to meet the requirements of the system. However, there are a growing number of dedicated integrated facilities management (FM) packages on the market which do not suffer from these limitations. However, these FM packages may not deal with critical maintenance functions such as condition surveys and repair orders to contractors in sufficient detail. Their use may therefore require additional bespoke or bought-in software packages for these functions, with the consequent problems of data format co-ordination and data exchange that the use of multiple systems would bring.

When considering an appropriate IT strategy, the maintenance manager is therefore faced with a number of issues:

(1) *Holistic or partial system.* Whether the IT system should seek to deal with all aspects of information, or part only. This decision will be influenced by the scale and complexity of the information flows within the maintenance organisation, and on the client organisation's IT strategy as a whole. A holistic system will have the considerable advantage of co-ordinating data throughout the maintenance operation. The attendant risks are those of disruption at changeover to the new system, and risk in terms of actual function. Additionally, it is more likely that a holistic system will require a bespoke design, with the attendant development costs and risks.

Partial systems seek to automate a particular function, such as the repairs logging and works ordering functions for day-to-day maintenance, or the condition survey function. For these packages there is a larger marketplace and therefore a wider choice from a range of products and providers. The risk of introduction is also much less, as teething troubles or even outright failure will only affect a particular range of operations rather than the organisation as a whole. However, there may be problems with the compatibility of data structures between a number of partial systems, making the interchange of data within the maintenance management organisation difficult without 'translation' software.

(2) *Bespoke or package.* The choice of a bespoke or package system will depend on the finance available, and the nature of the management organisation. Small or straightforward maintenance management operations may find most functions adequately catered for by an 'off the shelf' software system, particularly as these increasingly have scope for customisation of features to suit the user. Larger organisations are more likely to require a specially engineered package, which will require a long lead-in development and debugging time frame, and will cost much more not only on installation, but at intervals in the future when updating is required.

(3) *Scope and extent of system.* The range of functions available in IT software is extending. Previously unmanageable data, such as large image libraries, spatially referenced databases linked to CAD conventions, and the use of mobile communications, brings in many maintenance functions previously outside the scope of the information system. However with increasing scope comes increasing complexity of operation, and therefore the maintenance manager will need to decide the benefits of introducing new levels of IT within the context of efficient service provision. The existence of a capability does not necessarily point to the desirability of deployment, which should instead be measured according to the perceived benefits of using it. It may be decided to concentrate on alphanumeric data as being the cheapest and easiest to manage using well-tested IT, though it is increasingly likely that image and menu based data structure which relies on interactions with users via sophisticated visual environments, will shortly become expected rather than desired.

(4) *Future proofing.* Many maintenance organisations which have in recent years committed to introducing new IT have been disappointed that their IT solutions have quickly become obsolescent in the face of rapid IT developments. Software programs which may have taken years to develop (especially bespoke solutions constructed for a specific organisation) and which may then require extensive training of maintenance staff to operate, have become redundant in the space of a few years. A further problem relates to how data stored on older, obsolete systems can be interpreted by newer ones, though as software translation packages become more sophisticated, this problem may lessen. It follows that the introduction of IT systems in building maintenance management requires a long-term rather than a short-term view; with an evolutionary viewpoint being taken. This means the maintenance manager should look forward to developing capabilities to ensure that the specification for a replacement system may be

capable of upgrades in the future. As an example, the development of web-TV and the reduction in costs of digital cameras may revolutionise repair reporting by occupiers, who may be able to navigate 'smart' repair logging programs through their mobile phones or televisions using images as well as text, at any convenient time for them. In this context it is useful for the maintenance manager to examine other service sectors with better, more state-of-the-art IT provision, such as banking or on-line shopping, to get an idea of the coming capabilities of IT in the maintenance management sector.

6.8 The information system model

It will be apparent from the above that the crux of the use of IT in building maintenance management is in how it is introduced and implemented within an existing maintenance management organisation, to meet their specific needs. Whatever IT packages are introduced must be within a cohesive overall information structure within the maintenance management organisation. Failure to consider such a structure will lead to a lack of data co-ordination between different functions, the potential for breakdown in interdepartmental communications, and a tendency to take the data structures and methods offered by the computer software engineer (who may have no knowledge of maintenance management practice at all) rather than being able to interact with the software engineers in a meaningful way to arrive at a functionally satisfactory solution. For this reason, when assessing IT provision, it is necessary to consider the information system as a whole, and the efficiency of data flows within it. The broad categories of information in building maintenance management are illustrated in Fig. 6.9.

The information system model describes the maintenance management operation in terms of information flows. The methodology is that of information systems design and analysis. This field of study has been well developed for the purposes of IT software engineering and data modelling in computing, and is very well documented. The basic concept is the modelling of information as data pieces and the relationships between them. Its usual end-application is in the design of computer databases and software routines, but the general principles are applicable to the deconstruction and modelling of any large data-handling operation, such as a maintenance management information system. However, this must be preceded by the modelling and analysis of a 'human activity system' (after Checkland[14]) rather than attempting to model a database structure from the outset, since the database structure should be derived from the activities of the maintenance department rather than determine these activities.

The basic concept is that of modelling the maintenance management operation as a series of interrelating symbols (or data pieces). These data pieces may have many forms at many levels of complexity in reality (e.g. a dripping tap,

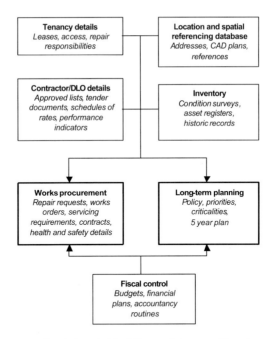

Fig. 6.9   Components of an integrated building maintenance IT system.

an irate tenant, a contractor's invoice), each of which only acquires meaning in its context (or its relationships with other data pieces). The three key structural concepts in this model are:

- *Semantics:* the meaning of the data (in terms of what it represents in reality)
- *Syntax:* the rules governing how the data pieces interrelate to provide meaning
- *Pragmatics:* how the data are perceived and used.

(after Benyon, D: *Information and Data Modelling*, 1990[15])

Information, in this concept, may be regarded as the data and its structure and interrelationships [15]. Thus the proposed system model for building maintenance information consists of:

- a definition of data used and its origin
- a description of the processes by which it is interrelated and structured.

As a simple illustration, we may have three data pieces, for example:

(1)   a want of repair such as a damaged door
(2)   a tenant who wants this repaired; and
(3)   the criteria for allocating the maintenance budget.

These can be linked by a process diagram to produce a *fourth* piece of data, i.e., its repair priority. The entire maintenance operation may thus be mapped as a series of interlinked data pieces as an information system model.

Information systems application to a maintenance management organisation

The first step in the process is to model the maintenance management system as an information system model, using standard conventions. These conventions entail the construction of flow diagrams which model the data, their inter-relationships, and the decision points that affect and are affected by the data. Because of the complexity of a large maintenance management organisation, there may be a number of such flow diagrams produced which model different sets and subsets of the maintenance process (Fig. 6.9).

The basic (conventional) information architecture for this modelling will take the form of modelling flows of data as inputs and outputs of a particular *process*. Each process 'box' may consist of a simple transformation of data (such as a comparison, say, between a want of repair, and the repair threshold to trigger an order being placed; or an addition, or of a decision), and will be dependent on the data inputs and the data outputs.

The model will be hierarchical inasmuch as high-level modelling of overall systems in the maintenance management organisation will result in the production of process 'boxes' as outlined above: these may be then regarded as the next lower level process in the hierarchy and themselves broken down into further process 'boxes', and so forth, down to the level of individual normalised data pieces (a normalised data piece is, broadly, a single 'fact' or, in computing terms, a byte of information; i.e., the smallest data piece required). However, as a result of the complexity of interrelationships between these data pieces (for example, where one data piece, say the property address, is used at different levels of the hierarchy, or in many different process 'boxes'), some aspects of relational database modelling will need to be adopted (after Bowers[16]).

Criteria for assessment

Following on from how the information system is modelled, the next methodological step concerns the criteria for assessment.

In maintenance management there is no ideal or absolute set of criteria against which the performance of a particular organisation can be measured[5], as a result of the diversity of circumstances in which each organisation operates. Whilst published sources on good practice are important for information and guidance, the use of such sources should be limited to informing the *breadth* of criteria which may be relevant for a particular organisation, rather than providing the *actual values* by which performance may be judged. For example, the RICS Guidance Note *Planned Building Maintenance* (RICS, 1990) delineates criteria such as preparing planned maintenance programmes, but gives no qualitative base against which actual performance can be measured.

It is possible to use the concept of customer service; i.e., how well the information system meets the expectations of the information system users and, ultimately, the building users themselves. The formal basis for measuring this may be the value management approach based on the work of the American engineer Lawrence Miles, who, writing in the 1970s[17], defined value analysis as: 'an organised approach to providing the necessary functions at the lowest cost.'

This lays out an accepted approach and framework for value analysis which has been successfully used in the building construction industry in recent years (Kelly & Male[18]), and which is applicable to building maintenance management. Kelly and Male make one particularly pertinent observation in the section on the second phase of the analysis process: information (the first being orientation):

'the information should not be based on assumption but be obtained from the best possible source and corroborated, if possible, by tangible evidence' (Ch. 2 p. 11)

The information can be analysed by means of a FAST analysis (functional analysis systems technique) whereby the operation of the maintenance department is broken down into a hierarchy of functional units. At the head of the hierarchy are the broad goals ('why'; equivalent to broad data classes such as 'tenancy details'), to the specific information required for a particular operation ('how'; equivalent to a particular repair detail such as a broken windowpane). Costs (of collection, processing, and outputs) are allocated to each of these units in the fast diagram; the costs of any one element being equivalent to the sum of all those elements in the hierarchical level below it. The pattern of information flow and its corresponding costs will therefore form the basis for the systems analysis required to design the databases and programs required to run these in the most efficient manner.

References

(1) Kochen, M. (1965) *Some Problems in Information Science.* Scarecrow Press, New York.
(2) Institute of Cost and Works Accountants (1967) *Management Information Systems and the Computer. Part 1: The Design of a Management Information System.* ICWA, London.
(3) Robertson, D. (1973) Data collection and analysis. Paper given at *National Building Maintenance Conference*, London, 1973.
(4) Hellyer, B. & Mayer, P. (1994) *Stock Condition Surveys – A Basic Guide for Housing Associations.* National Federation of Housing Associations, London.

(5) Wordsworth, P. (1992) *Standard Maintenance Descriptions: Part 1: Maintenance Management Practice Paper No. 20.1* Royal Institution of Chartered Surveyors Publications, London.

(6) Ministry of Housing and Local Government (1964) R*eport of the Working Party on the Costing of Management and Maintenance of Local Authority Housing.* HMSO, London.

(7) Damen, T. (1988) Meaningless and Meaningful Maintenance Planning. Paper given at *CIB/W70 Whole Life Property Asset Management*, Edinburgh, 1988.

(8) Vickery, B.C. (1965) *On Retrieval System Theory.* Butterworth, London.

(9) Bushell, R. J. (1970) Building planned preventive maintenance. *Building Maintenance.* Sept. 2–9.

(10) Building Maintenance Information (1997) Computer Aided Maintenance and Facilities Management Systems for Building Maintenance. *BMI Special Report No 260* June 1997, BMI, Kingston.

(11) Spedding, A. (1990) Management of Maintenance – The Need for and Uses of Data. Paper given at *CIB/W90 Building Asset Management, Chartered Institute of Building*, Sydney, 1990.

(12) Wordsworth, P. & Boughey, J. (1993) Information Needs and Information Technology in Management. *Property Management Journal* Vol. 11 No. 4 Autumn 1993.

(13) Allen, S. & Hinks, J. (1994) Information Technology and Effective Maintenance Management: A Methodology to Identify the Key Variables in this Relationship. Paper given at *ARCOM 10th annual Conference*, Loughborough, 1994.

(14) Checkland, P. (1981) *Systems Thinking and Systems Practice.* John Wiley and Sons, London.

(15) Benyon, D. (1990) *Information and Data Modelling.* Blackwell Science, Oxford.

(16) Bowers, D.S. (1988) *From Data to Database.* Van Nostrand Reinhold, London.

(17) Miles, L.D. (1972) *Techniques for Value Analysis and Engineering.* McGraw Hill, New York.

(18) Kelly, J. & Male, S. (1993) *Value Management in Design & Construction.* E. & F.N. Spon, London.

Chapter 7
The Maintenance Organisation

7.1 Organisation theory

It is rare that one has the opportunity to plan the structure of an organisation 'de novo'. Usually the maintenance manager takes over an existing structure that has evolved to its present form over a number of years. Possibly, the most important factor has been the personalities of the key members of staff. It is helpful, therefore, to understand the pressures and influences which have determined the nature of its development, and of any constraints which may restrict future changes. There is no single structure that is equally appropriate for all organisations. The most appropriate is that which is best suited to meet the particular needs of the organisation.

A great deal has been written about the theory and practice of management and it is only possible in this book to give an outline of the main lines of thought on the subject. The early studies were concerned primarily with the individual worker, then attention was turned to the organisation as a whole and the behaviour of groups of workers and more recently to the total system in which the organisation operates. F. W. Taylor, an American engineer, is generally credited with being the originator of scientific management. He realised the advantages to be gained by a systematic study of the way in which repetitive operations were carried out. He found that complicated jobs could be broken down into small units enabling workers to specialise in only a part of the whole job and thereby become more proficient and productive. He assumed that workers were motivated primarily by cash incentives and that the simplification of job method would not only benefit management by the resulting higher productivity but would also lead to higher pay for the workers and be mutually advantageous. F. Gilbreth, a building contractor, applied motion study to bricklaying and after careful study of the body movements involved in laying a brick devised ways of cutting out unnecessary movements so that a brick could be laid with only five motions instead of the eighteen required using the old method. H. Fayol, a French mining engineer, also applied Taylor's ideas and from them developed a set of general principles covering such aspects of management as division of work, authority, responsibility and discipline. However, others were beginning to criticise Taylor's basic notion that workers are motivated solely by cash incentives. Mary Parker Follet developed the 'law of the situation' which saw management as a series of changing situations each dependent on and conditioned by the preceding situation. She argued that orders should not be given because 'I say so' but because the situation demands it. Thus, managers are as much under orders as the workers in that both must obey the law of the situation. The simplistic views inherent in Taylorism were further challenged by the Hawthorn experiments. These were a series of studies carried out by Elton Mayo and others at the Hawthorn plant of Western Electric in America. Initially, a group of workers was tested to ascertain the effect of various levels of lighting on their rate of working. It was found that there was a general improvement as the lighting levels were progressively increased but that the rates continued to improve when the electricians only pretended to fit brighter bulbs. One explanation of this so-called Hawthorn effect was that the operators, being aware that they were being tested, displayed preconditioned responses to the situation, i.e., that the very act of monitoring the operators' performance influenced their behaviour. These studies provided a valuable insight into the behaviour of working groups and led to the 'human relations' movement that placed greater emphasis on the social environment of the workers. It was recognised that a work group had a particular personality which was distinct from that of its individual members. Also that informal groups, i.e., those formed by casual meetings, say, in the canteen, are just as important as the formal groups which have been selected and allocated to a joint task by management.

More recent studies have been concerned with management as a socio-technical system. Handy[1] uses the term 'cultures' to describe the norms or values that determine the way in which an organisation is operated and structured. He classifies these cultures as:

(1) *The power culture* which depends on a central power source. This is typical of small family businesses and entrepreneurial organisations that have few rules and procedures and which are highly dependent on the person or persons at the centre. They are quick to react to change.
(2) *The role culture* in which roles and job descriptions are precisely specified. This is typical of local authorities and other bureaucratic organisations and is characterised by the laying down of job descriptions, limits of authority, procedures for communications, etc. It works well in a stable environment but is slow to react to change.
(3) *The task culture* in which people and other resources are organised to achieve some set task. This is essentially a team culture in which individuals have a high degree of control over their work. It is useful for new ventures but difficult to exercise control except through allocation of resources.
(4) *The person culture* in which the individual is the central point and the organisation exists merely to serve the interests of its members, e.g. barristers' chambers. Certain specialists in an organisation may also regard it as an opportunity to do their own thing and owe little allegiance to the organisation.

There may be different cultures in the same organisation, e.g. the main culture may be a role culture to deal with routine maintenance work but with a research and development section working on a task culture.

It is now popular to adopt what is called a systems approach. This stems from the work that was originally carried out by N. Wiener[2] who viewed an organisation as an adaptive system that is entirely dependent on measurement and correction through information feedback. The basic idea is illustrated in Fig. 7.1.

The model may be extended to take into account all the social and economic pressures that affect the work group as illustrated in Fig. 7.2.

The move towards a more open systems approach has led to the contingency or situational approach. The contingency approach means that different environments require different organisations and that there is no one best type of organisation. For example, it means that job enrichment should be applied with the realisation that some operatives do not want their jobs enriched. Each person and situation is different and problems are created by the adoption of a single value system. Early evidence of this was provided by J. Woodward who showed that the form of organisation was related to the technology used, which she classified as unit production (small batch), mass production (large batch) and process production. This has been followed by studies by Lawrence and Lorsch[3] who have shown that stable conditions favour the classical forms of organisation whilst in changing conditions the opposite is true.

7.2 Functions

The term 'maintenance organisation' is used in this context to describe the person or persons responsible for the planning and control of maintenance operations. In a small firm, the functions may be undertaken by a member of staff in addition to his other duties, while in a large firm there would usually be a separate group of people solely responsible for maintenance. The organisation would also include independent consultants who are called in from time to time to advise on particular problems. Whatever the scale, the basic functions are broadly similar and include the following.

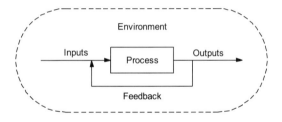

Fig. 7.1   Basic information feedback loop.

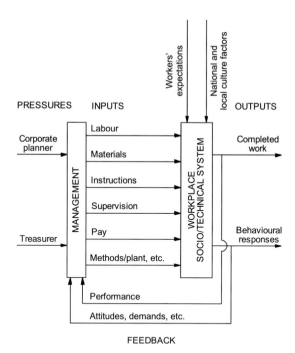

Fig. 7.2 Enhanced information feedback loop.

Advisory function

This would involve liaison with owners and users and consultation with upper management on such matters as:

(1) The standards to be maintained and the effect on user activities of deviations from these standards.
(2) The relative merits of alternative maintenance policies and the extent to which it would be advantageous to employ operatives directly for executing the work.
(3) Clarification of any constraints in relation to limits of expenditure, desirable cash flow patterns, acceptable delay times or restrictions on time and method of carrying out work.
(4) Estimates of maintenance expenditure both long and short term, including, where appropriate, the cost of initially bringing up to the required standard and the possibility of phasing any such backlog over a period of years.
(5) Provision of cost and other data to assist upper management in deciding whether to repair or renew.
(6) Technical requirements for minor works involving alterations or small additions to the building; although not strictly maintenance, it is usual for the maintenance organisation to assume responsibility for this type of work.

(7) Advice on the maintenance implications of designs for proposed new buildings. It is most important that the maintenance organisation should participate in the briefing of the designer and be given the opportunity to comment on the detailed design.

Organisational function

This may be in relation to the central administrative and supervisory system or to the execution system whether by direct labour or contract.

(1) *Central administrative system.* These functions would be necessary and the associated costs incurred whether the work is ultimately let to outside contractors or undertaken by directly employed operatives.
  • defining the duties and responsibilities of administrators and supervisors and of their technical and clerical supporting staff
  • establishing job relationships, patterns of accountability and paths of contact
  • formulating standard procedures and operating instructions
  • devising an appropriate information system and channels of communication to ensure co-ordination and effective feedback for control purposes
  • provision of suitable office accommodation and equipment
(2) *Direct labour force.* Where the work is undertaken by an independent contractor, the contractor would assume the responsibilities in this area.
  • selection of supervisors and definition of their duties and limits of authority; alternatively, the functions of the supervisors may be considered in the context of the central administrative system
  • engaging operatives, including payment, timekeeping procedures, administration of incentive schemes, provisions for safety, welfare and training
  • purchasing materials and stores control procedures
  • provision of plant either by purchase or hire
  • arranging for the transport of labour, materials and plant to and from the site
  • upkeep and maintenance of workshop and stores buildings
(3) *Contract work.*
  • preparation of tender documents and selection of a contractor
  • administration of the conditions of contract, including authorising and valuing variations and certifying for payment and satisfaction
  • supervision of work to ensure compliance with terms of contract

Control

The control functions are dependent on the timely receipt of accurate information relating to the state of the system. The control functions operate in the following areas.

(1) *Work input.* Identifying the extent of work necessary to achieve the required standards within the constraints laid down. The processes involved would include planned inspections, appraisal of user requests and assignment of priorities.

(2) *Time of execution.* Programming the workload so that the carrying out of the work is timed in accordance with the needs of the user and the available labour force. This would normally involve the preparation of long-term strategic plans and of short-term tactical plans when the requirements are known with greater certainty.

(3) *Quality.* Supervision of work during execution and by subsequent control inspections to detect latent defects.

(4) *Cost.* Budgetary control system including estimating resource requirements in cost and performance terms for later comparison with actual cost and performance achieved.

(5) *Feedback.* This is an inherent feature of all the control functions and would involve keeping such essential records as are necessary for the proper control of the operations.

Miscellaneous functions

In addition to the basic functions described above, the maintenance organisation may have responsibility for various other matters such as:

- safety and security, principally in relation to compliance with statutory fire precautions and the proper maintenance of fire fighting equipment
- refuse disposal
- cleaning
- grounds, etc.

7.3 Structure

The structure of an organisation is usually represented by a chart showing the allocation of formal responsibilities and the linking mechanisms between the roles, i.e., the typical line and staff chart. This is usually backed up with a corporate plan setting out the general obligations and policies of the organisation and job specifications which outline the duties of the various members of staff and indicate to whom they are responsible and the limits of their authority. The types of structure appropriate to maintenance may be described as centralised (or functional) and decentralised (or territorial or geographical) as illustrated in Figs 7.3 and 7.4. The extent to which power is retained by the central organisation will depend upon the advantages to be gained from standardisation and the adoption of common procedures centrally administered and controlled.

The needs of different firms are so varied that there is clearly no one type of organisation that would be equally successful in all cases. A survey of a small sample of firms failed to reveal any uniformity of practice. In many cases, it was apparent that where the firm had expanded and acquired additional premises, the maintenance organisation merely increased in size without any change in its original character, in spite of the different circumstances under which it was operating. Much depended upon the personalities of the particular individuals concerned.

The size and structure of the central organisation will depend upon the following factors:

(1) The volume of the workload will determine the staff time required for inspections, estimating requirements, preparation of drawings and technical documents, programming and control functions.

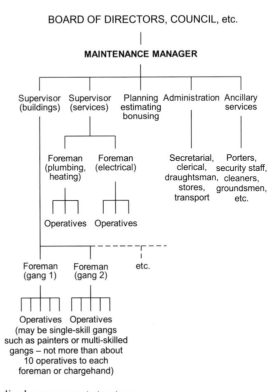

Fig. 7.3 Centralised management structure.

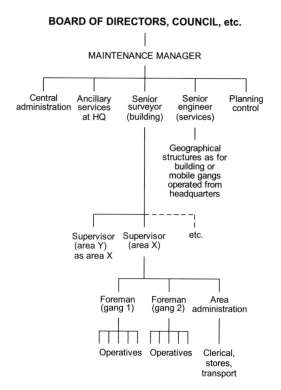

BOARD OF DIRECTORS, COUNCIL, etc.

MAINTENANCE MANAGER

Central administration | Ancillary services at HQ | Senior surveyor (building) | Senior engineer (services) | Planning control

Geographical structures as for building or mobile gangs operated from headquarters

Supervisor (area Y) as area X | Supervisor (area X) | etc.

Foreman (gang 1) | Foreman (gang 2) | Area administration

Operatives | Operatives | Clerical, stores, transport

Fig. 7.4   Decentralised management structure.

(2)   The nature and complexity of the work and whether predominantly building or engineering will determine the desirable qualifications of the supervisory staff.

(3)   The location and dispersal of the work will influence the travelling time and hence the number of supervisors required to maintain effective control.

(4)   The timing of the work and, in particular, the need for certain work to be habitually undertaken outside normal working hours may demand some duplication of supervisory staff to ensure continuous control.

(5)   The skill and reliability of operatives will determine the amount of information that they require and the frequency of visits to check progress and maintain quality control.

(6)   The method of executing the work, i.e., whether by direct labour or by contract, will have obvious repercussions on the organisational structure of the maintenance department.

(7)   The responsibility of the maintenance department for minor new works or miscellaneous services.

(8)   Building owners' or users' policies for different types of buildings.

In practice, one finds a wide range of structures ranging from the extremely informal arrangements common in small private firms, to the complex hierarchical systems adopted by large local authorities and government departments. A comparison of these extremes will serve to illustrate their relative merits and shortcomings.

In the small firm the proprietor may work alongside the staff and discuss with them not only job information but also feelings, values and standards. In this type of organisation, the problems are usually only roughly categorised and the operatives have only a few loosely defined routines. The structure has few levels that are not clearly distinguished and co-ordination is achieved by the overlapping and intermingling of roles. It is extremely adaptable and well suited for dealing with the uncertainties of maintenance. However, it does require competent operatives and a stable working relationship between the members of the group. It is particularly advantageous where the firm employs highly skilled operatives who specialise in certain aspects of maintenance.

On the other hand, the organisation of a large direct labour force does not permit the same degree of participation by the operatives. The work is usually divided into a number of small, clearly defined units that are managed by formalised routines. The relations between these units are laid down in further routines with detailed job descriptions and rules and procedures for each man. Co-ordination between the units is achieved by a hierarchy of positions with the powers of each clearly defined. The principal defect is the lack of flexibility and inability to deal with conditions that were not foreseen when framing the rules and procedures. It works best where there is a repetition of the same problem for which the method of working can be clearly defined. It does not, therefore, require the same degree of expertise on the part of the operative and there is a restricted scope for the exercise of discretion.

Also, the large complex organisations demand a level of management expertise not readily available in this area. The Emmerson Report[4] noted the lack of people with appropriate management skills throughout the construction industry and that those who have the necessary qualifications are attracted more to new work than to maintenance. This point was emphasised in the Woodbine Parish Report in relation to maintenance supervisors in the hospital service. Of the 240 or so building supervisors employed by the service in 1969, 85 held no technical qualifications and only 63 had a Higher National Certificate or higher qualification. It was reported by the hospital authorities that those with the higher qualifications did in fact achieve better standards of maintenance and improved organisation.

A report by the Construction Industry Training Board[5] drew attention to the serious weaknesses in the management structures of small and medium-sized building firms. To combat such problems, the report recommended broadening the educational preparation for supervisory and management roles. However, it would seem that the situation can be improved only by recognition of the importance of maintenance and reassessment of the status and rewards due to persons responsible for the organisation of this work. In the case of local authorities, it has been noted[6] that the organisation of maintenance may form part of the duties of the engineer, the architect or the housing manager, but only rarely is there a building manager with sole responsibility for planning the execution of this work. Because of the routine nature of such maintenance work, authorities are inclined to leave effective control in the hands of officers who regard it as secondary to their main duties and are not equipped to consider the longer-term issues involved.

Management at every level develops its own policy within its own sphere of responsibility and scope for decision making. Broadly, policy is to do with the setting of objectives and the means of achieving them. Decisions at the different levels in the organisation must be compatible and clearly understood and implemented by all who are directly concerned; others who may be affected but who are not directly implicated must be informed. Clearly, success is dependent upon the ability to set and achieve the best objectives at all levels. This has been described as 'management by objectives'. It should not be overlooked that people work more effectively together if their individual objectives are parallel and do not diverge. A major part of labour relations is to do with keeping the objectives of all concerned sufficiently close to ensure that the objectives of each are achieved by the joint efforts of all. 'Management by exception' is another important concept in this context. It means that by having established objectives and the means of achieving them attention can be concentrated on deviations or 'exceptions' from the main plan. This involves providing adequate means for monitoring performance against targets set, so that exceptions can be identified sufficiently early for corrective action to be taken.

7.4 In-house or outsourced staff

The choice between employing consultants or tradespeople directly, or engaging an independent contractor for that purpose, should be decided according to which offers the greater advantage in terms of cost, quality and convenience. However in practice this choice may be heavily influenced by outside forces such as political directions or overall company policy. Usually a large organisation will use both methods, the problem then being to decide the amount and type of work that should be carried out by each. Potentially, in-house staff are more economic than outsourced staff by at least the profit margin included by the contractor in his tender. However, achieving this saving demands an equivalence of management skills so that the performance of the in-house operative is not inferior to that of the operative employed by the contractor. There are other factors to consider besides direct or tender cost, in particular the cost of carrying the staff overhead in terms of payroll, tax, redundancy obligations, etc., and a decision must rest on an analysis of the particular advantages and disadvantages in the context of the needs of the organisation.

Main advantages derived from in-house staff

(1) Full control of the allocation of work to operatives, resulting in a quicker response to emergencies and greater flexibility in terms of the work done and the times worked. Where the work is carried out by a contractor or consultant the instructions must pass through the contractor/consultant who may be unable to withdraw staff from other jobs as quickly as the situation demands. Losses flowing from the interruption of the firm's activities during these periods of delay should be taken into account when comparing the costs of in-house and outsourced staff.

(2) In-house operatives acquire an intimate knowledge of the buildings and become familiar with the user's requirements and any constraints on the execution of the work. This simplifies the communication of job information in that the operative has a background knowledge of the situation and is able to understand what is required more easily and with less risk of misunderstandings. It also engenders a better working relationship between the user departments and the maintenance department.

(3) Better control of quality through the employment of people of known and tested ability and through direct supervision. The in-house operative will have a greater sense of identity with the firm and will see maintenance as a continuing process in which defective work will only create greater problems at a later date. There is thus a greater incentive for the in-house operative to do the work properly the first time. However, whether or not this advantage is realised will depend upon the policy of the user organisation. In a report on local authority incentive schemes it was noted that it was the policy of a number of councils to give employment to people who were otherwise unemployable. While it might be socially advantageous to provide welfare in this way, it results in maintenance costs being higher than they need be and prevents the comparison of direct labour and contract on the same basis.

It is important to consider the total cost of maintenance over a period of time rather than marginal savings on an individual job. Thus there is nothing to be gained in the long run by skimping the preparatory work for external painting in order to reduce costs. The result will be a reduction in the time of the painting frequency and a need for more extensive work at the next repainting period. For such work it is generally thought that direct labour ensures a higher quality and a longer lasting job.

(4) By employing operatives directly it is possible to gain a first-hand knowledge of the factors which influence output and thereby to develop more effective cost control procedures linking planning and execution. It is also possible to carry out controlled experiments with different methods in order to achieve the same standard at a lower cost. Usually, where work is let out on contract, only the lump sum price is recorded for feedback

purposes. This is of little value for estimating and programming future work, for which work-hours would be more useful.

(5) Delay is not incurred in inviting tenders or negotiating with contractors, and the procedures for initiating the execution of work are simplified.

(6) Where the work involves a security risk it is advantageous to have knowledge of the background of the operatives.

Disadvantages attributed to in-house staff and operatives

(1) There is a lack of specialisation in terms of labour skills and plant. However, this criticism could be applied equally to the small building firms that execute the major part of maintenance work. Many small property owners make great use of the jobbing builder who employs a few handymen and calls upon self-employed tradesmen where necessary. On the other hand, the larger direct labour forces are demonstrably better equipped than most of the very small, severely undercapitalised independent contracting firms. The assertion is really only valid in respect of certain types of work, e.g. remedial treatment to timber, painting, cleaning, etc., in which specialist firms have developed a particular expertise. Similarly, a maintenance organisation may choose to use consultants where there are specialist needs such as structural engineers or surveyors for large condition surveys that may warrant the importation of in-depth specialist skills not available from within the organisation.

(2) There is a feeling in some quarters that direct labour or in-house staff are more expensive than contracted-in staff. Indeed, this is the most common complaint levied against direct labour forces employed by public bodies. It was for this reason that the Local Government Planning and Land Act 1980, required local authority direct labour organisations to tender in competition with contractors for works over a certain value and to show a prescribed rate of return on capital involved. However, comparing tenders is not a valid way of measuring either the efficiency or the usefulness of a direct labour force for the following reasons:

- The submission of the lowest tender does not automatically indicate that the contractor submitting that tender is the most efficient of those tendering. A low tender may be the result of inaccurate estimating or of a shortage of work in the locality, or it may be set deliberately low with the intention of cutting the quality. The true measure of efficiency, as far as the contractor is concerned, is the margin of profit; clearly there is no advantage in submitting a low tender if the result is a net loss. Direct labour forces do not make a profit and therefore must measure efficiency in other ways. The only true comparison would be of basic costs of labour and material expended on similar jobs, but as no two jobs are identical the result would be distorted by differences in the nature of the work and the conditions under which it has to be executed.

- A contractor or consultant is free to bid for any work available and, not being restricted to a particular client, may select that which presents the fewest difficulties. On the other hand, the direct labour force is confined in its operations and is frequently placed in the position of having to carry out work which by its smallness or complexity is not attractive to contractors or consultants. In such cases either it is not possible to find a contractor who is willing to do the work or the price quoted is excessively high.

- Continuity of work within an overall plan is one of the biggest economic advantages to be gained from the use of a direct labour force. It is, therefore, self-defeating to require it to justify its efficiency on each and every individual project by tendering in open competition with outside contractors. The result would be to introduce uncertainty in the planning that could disrupt the even flow of work programmed for the force, resulting in idle time and under-utilisation of plant and inhibiting the forward bulk purchasing of materials.

- There are inevitably many small, time-consuming jobs that usually, either because of size or urgency, fall to the lot of the direct labour force. In a hospital memorandum[7] it was stated that up to 50% of the total maintenance budget was absorbed in dealing with minor breakdowns and *ad hoc* requests. Although it was anticipated that this proportion would be considerably reduced by the introduction of a planned maintenance system, the need for a 'fire brigade' service could not be eliminated altogether.

- The overall efficiency of a direct labour force cannot be judged on the basis of the result of an individual competition. A direct labour force is not able to balance gains on one contract against losses on another and the single criterion of cost can be misleading unless it takes into account all the benefits.

(3) A further disadvantage that is sometimes pointed out is that the operatives may be at times underemployed. This may be inevitable if the circumstances demand an extremely short response time. In such cases, efficiency should be measured by the average time between failure and rectification as well as the cost of providing the required service. However, this sort of situation would merit close investigation to discover whether the emergency calls can be reduced by increasing the level of preventive maintenance. Filling the time with work of doubtful value is, of course, something that must be avoided.

(4) It should not be overlooked that the use of direct labour and in-house staff necessitates the provision of supporting facilities in the form of building space for offices, stores and workshops as well as additional administrative and clerical staff to deal with work control, incentive schemes, labour relations, payroll and accounting services.

(5)   A further point to take into consideration is that the financial and other risks are borne solely by the organisation employing in-house staff. Where work is let to outside contractors on a lump sum basis the extra costs occasioned by unforeseen circumstances are usually the contractor's responsibility and, except as provided in the contract, cannot be passed on to the client. Also the contract normally contains safeguards which entitle the client organisation to redress if the work is not carried out in accordance with instructions or is not completed on time. However, many of the smaller jobs are let on a cost reimbursement basis and this throws the financial risk on to the client who is at the disadvantage of not being able to control directly the job method.

In respect of nonmanual services in building maintenance management, the decision to use consultants for some types of administrative and professional work will be related to the nature of the organisation, and the rate of change in it. A static administrative and professional cadre of staff can be a positive benefit during a stable business climate, but may become a liability in times of rapid change, because of the costs of retraining and redundancy which may be entailed. Outsourcing strategies may be demanded at a higher corporate level than the maintenance manager, whatever the practical benefits of having in-house or consultancy staff, because of the need to keep cost overheads low and thus stock earning/cost gearing more attractive to investors.

Over the last twenty years this has led to a significant trend towards 'flatter' management structures within client organisations, in which there are fewer hierarchical layers and which concentrate on the core business of the organisation. All needs peripheral to the core business (which often may include the building maintenance management role as well as the maintenance procurement function) are contracted out on the basis of fixed price contracts, term contracts, or service agreements (see Chapter 9). Park[8], in his book *Facilities Management: An Explanation,* looks at this question from the wider remit of the facilities manager, and expresses the point succinctly: 'There are advantages and disadvantages in having a maintenance department, the main advantages being continuity of staff with benefits in security terms. The disadvantages when compared with engaging a term contractor are lack of flexibility and management involvement'. Thus the division between the 'client' organisations and those responsible for executing the work has become more pronounced. Even within large organisations (both public and private) which contain both the client function and the maintenance operatives, internal division into self-standing 'business units' has resulted in a situation where outsourcing by formal contract, or by quasicontractual arrangements within the organisation, have become the norm.

The building industry has seen similar changes to the aforementioned compartmentalisation of management within client, or property-owning organisations. Most local authority direct labour organisations are now free-standing business units in their own right, with no secure or guaranteed workload from the authority. Many building contractors have themselves moved towards an in-house/outsourced model by which a cadre of permanent in-house (direct labour) operatives are retained, and particular peaks in demand met by taking on temporary workers either directly, or by contracting out the work to subcontractors. For example a contractor winning a two-year measured-term contract may require a larger labour force just in order to service this contract, and therefore has the choice of engaging permanent direct labour, or contracting out to individual tradesmen or other subcontractors. The issue for both the DLO and the contractor is what proportion of directly-employed operatives should they keep as their established workforce, and what proportion should be outsourced. There are a number of factors influencing this choice.

Nature of work

The programmed work should be divided according to the skills required for its execution, i.e., traditional craft skills, specialist skills and relatively unskilled or semiskilled work.

- Work requiring conventional craft skills would normally be the main commitment of a direct labour force. In certain cases the firm may employ skilled craftspeople in other capacities, e.g. shop fitters, who are able to cope with the day-to-day requirements of building or services maintenance.
- Specialist work is usually let to outside contractors since there would rarely be sufficient work to provide continuity for a permanent labour force. Also a specialist firm would have a better knowledge of the characteristics of the component and possess the necessary tools and spare parts to effect a repair more quickly. In some cases the appointment of a specialist firm may be dictated by safety considerations, e.g. electrical installations, lifts, etc.

Volume of work

The total amount of work in each of the above categories for each trade (or compatible combination of trades) should be assessed and annual and seasonal variations identified. From this it will be possible to determine:

- Whether or not the amount of work of one type will provide full and continuous employment for at least one person in each craft group. The number of operatives can be obtained from the estimated costs included in the long-term programme by assuming an average ratio between total cost and the work-hours included therein.
- The work which should be let to outside contractors to deal with peaks of activity beyond the capacity of the direct labour force. For this the workload should be analysed into:
- routine day-to-day and frequently recurring minor works
- major works divided into 'cyclic' and 'infrequent'

An outline programme for the budget period taking into account any restrictions on the timing of the work imposed by the user of the building will reveal the peak periods when outside assistance will be required.

Response time

The workload should be analysed according to the degree of urgency of the work sections so that an assessment can be made of the response time necessary to avoid consequential losses or inconvenience. A breakdown into three categories would be adequate.

- *Emergency work* – the breakdown of vital services or fabric defects involving risk to the occupants or contents. Clearly this work must be dealt with promptly, preferably on the same day. However, it is rare that breakdowns calling for emergency repairs are completely unpredictable. Procedures should be laid down for initiating and executing such work outside normal working hours. In many cases, to be effective, speed is essential and this would favour direct labour or special arrangements with a contractor.
- *Urgent work* – defects, which while not a positive danger, cause some inconvenience to the user and may develop into more serious faults if left unattended. An acceptable response time would be one week and this again would tend to favour the quicker initiation of work possible with direct labour.
- *Normal work* – work which while necessary and desirable does not affect the immediate user of the building and which may be phased in with the annual programme. Whether it is carried out by direct labour or contract will depend very largely on its magnitude.

Location

Different views have been expressed on the relative merits of direct labour and contract for the maintenance of buildings in remote areas. In some cases the circumstances may favour the direct employment of labour in areas where the use of contract labour is difficult by reason of remoteness. In other cases it may be found more convenient to employ local contractors or tradesmen on an 'on call' basis, the object being to avoid excessive travelling time caused by calling direct labour from a central unit. This difficulty could be overcome by decentralising the organisation of the direct labour force if the individual buildings or groups of buildings provide a continuous workload for a resident maintenance team. The central workshop would provide a pool from which specialist staff and equipment could be sent to peripheral units should the occasion demand: a facility not available to the small local builder.

Quality

The importance of maintaining appropriate standards must be considered. The main contractor or DLO will be directly responsible for the quality of its workforce, whether outsourced or directly employed. The surest way of obtaining such standards is by the direct employment of highly skilled operatives who can be more closely supervised than contract labour.

Security

The need for special security precautions may be the dominant factor overriding the probability of any marginal cost advantage. This might preclude the presence in the building of a contractor's casual labourer who has no responsibility to the firm.

Availability of space

The employment of direct labour necessitates the provision of supporting facilities such as stores and possibly workshops. The question arises therefore as to whether it is physically possible to provide the space or whether an alternative use of the available space would be more profitable.

Market conditions

The decision to employ a direct labour force is essentially a long-term one and therefore an assessment should be made of the probable future differences in cost as well as those that exist at the present time. It is clear that the prices charged by contractors will be influenced to a very much greater degree by market conditions than will the costs of direct labour. In addition, it is probable that in times of severe labour shortage the greater permanence and security of direct employment will be regarded by the older and more mature tradesmen as being preferable to higher wages for an uncertain period.

Market conditions vary according to time and place. The time-related changes may be long-term and progressive or short-term and fluctuating. The short-term changes may be of a predictable seasonal character or unpredictable and isolated. Different locations may follow a broad regional pattern and depart from the general trend in a consistent way or may be highly individual. The inconsistencies may arise from various causes: for example, the closure of a major industrial plant may result in a local depression with consequential effects on maintenance and other prices. Another factor that impinges directly on maintenance is the volume of new construction in the locality. Most small and medium-sized builders carry out new construction in addition to maintenance and usually prefer new work because it is more straightforward and offers potentially higher profits. Any curtailment of the new construction programme either as a result of the general economic climate or deliberate government policy will divert these firms to the maintenance and improvement market, thereby increasing the competition in this field with a likelihood of lower prices.

Cash flow

A further factor is that the larger part of maintenance expenditure is attributable to labour, and where operatives are employed directly wages must be paid at weekly intervals. Payment for contract work is not made until after completion of smaller jobs, or at monthly intervals or predetermined stages for larger jobs. In addition, it is normal to hold an agreed percentage of the amount due to the contractor until the end of the defects liability period. Of course, the cost of providing working capital should be included in the contractor's tender, but this may well be less than the profit which could be obtained from an alternative use of the finance by the client organisation.

Consideration of the foregoing factors will enable work to be apportioned between direct labour and contract in a manner which will best serve the interests of the organisation.

There are various compromise solutions designed to blend the best features of direct labour and contract. For example, an arrangement can be made with a firm of contractors to keep a mixed gang continuously on the site under a resident supervisor. The instructions would be given directly to the supervisor by the maintenance supervisor and the work paid for on a schedule or cost-plus basis. The main advantages obtained from this system are that the staff are permanently on the site although the number can be varied according to the workload. The client organisation avoids the responsibility for employing labour in regard to wages, pensions, holiday payments, trade union negotiations, etc., and any special equipment can be obtained quickly from the contractor. Also the contractor can buy common materials in bulk and obtain more favourable terms from manufacturers and merchants. It is probable that the continuity of work resulting from this arrangement would encourage the contractor to quote lower prices.

Total costs

Where an organisation uses both direct labour and contractors the object should be to minimise the total costs of executing maintenance work as represented in the following equation:

$$T = (C + S) + (L + M + P + S) + A$$

where:
T = total costs
C = contractors' costs
S = supervision costs
L = labour costs
M = materials costs $\quad\Big\}\quad$ direct labour
P = plant costs
A = administrative costs

7.5 Works order system

Whether the client organisation's maintenance procurement is through direct labour or contract, there will need to be an effective system in place to procure maintenance works. For fixed-price contracts such as rewiring and roof replacements, a formal contract and specification will be used. The format and structure for this type of procurement are dealt with in Chapter 8. For day-to-day works, and for larger works procured under a schedule of rates, a works order system is an essential component of the maintenance organisation. Repair requests and needs will be taken, scheduled, and prioritised by the client maintenance management organisation. The works order system is the process whereby these orders are put out to the contractors, and by which the contractor or direct labour operative accounts for works done. Under a measured term contract the works will need to be specified and charged in accordance with the priced schedule, which essentially consists of fixed rates for items of labour (usually in the form of a trades and hours worked classification) and materials used.

The contractor will invoice the client organisation on the basis of the schedule of rates, and its provisions for extras, overheads, etc. It is usual for the client organisation to scrutinise these invoices for accuracy prior to payment. Obviously, it will not be efficient for a supervisor from the client organisation to make arrangements to visit the site of every such repair, or else the cost of supervision would become a significant percentage of the total cost of the repairs, making them uneconomic. It is therefore more usual to inspect a given percentage of invoices submitted. This may range from 3–5% in the case of straightforward repairs by trusted contractors, to 10% in cases where the work is complex, or the contractor and/or operatives require a degree of audit to ensure they are accounting for work done and materials used in a fair and accurate way. Any significant discrepancies would then trigger a wider audit of work done, with possible sanctions up to and including repudiation of contract depending on the seriousness of the errors.

The contractor will need to process the client works order in somewhat greater detail in order to arrive at an accurate breakdown of time and costs, and to be able to track levels of stocks and building materials held, to enable reordering, and labour time planning. In this respect the key document used for controlling work input is the work order (WO) or Job Card. It provides a medium for the recording of work-hours and materials that can subsequently be analysed for various management purposes. There are in use many different types of forms of varying degrees of complexity, but the basic information they should give is:

Number and date of issue

Location codes
Codes to indicate location and other classification categories into which costs are required to be analysed.

Priority for unscheduled work

The priority will depend on the effect of the state of disrepair on the user of the building and in particular on safety and health considerations. It is difficult to make fine distinctions, but in practice three categories should be sufficient:

- Priority I: emergency work to be carried out on the same day.
- Priority II: urgent work which while not constituting an immediate danger should be carried out within a week of notification.
- Priority III: normal work that would account for the major part of the budget allowance for contingency maintenance and may be programmed on the same basis as the scheduled work.

Description of work

As the work orders are the main source of information on costs for subsequent estimates the descriptions should be standardised, i.e., identical descriptions should be used for repetitive jobs. Usually the descriptions are too vague to identify precisely the work content, the presumption being that the details will be transmitted verbally on site. This not only leads to delays on site, but also renders the work order virtually useless for estimating purposes. A schedule of standard descriptions can be taken from the schedule of rates, to facilitate the storage, retrieval and processing of cost data.

However, standard descriptions may not indicate the particular conditions under which the work has to be carried out and therefore, if an unusual method or sequence of operations has to be adopted, these should be noted so that future comparisons are not distorted.

Estimated work-hours

The estimated labour content expressed in work-hours or work-days should be entered on the order or, where a bonus scheme is in operation, the target time. Quite apart from the control value of comparing actual against estimated times, it focuses attention on the job method and the need for proper programming.

Materials

The quantities of each material should be stated in sufficient detail for stores requisitioning or purchase.

Actual work-hours

The actual work-hours on the job are required for cost control and calculating bonuses. However, as they will form the basis for future estimates, a distinction should be made between productive and unproductive time and, where the latter is excessive, the reasons why this was so should be noted. It is also necessary to distinguish between emergency work and that of a lower priority, since this will govern the extent to which jobs can be grouped to reduce travelling time and hence costs. This will make the additional cost of providing an emergency service explicit for comparison with the benefits obtained by way of extra safety and convenience.

All work carried out should be covered by an appropriate work order. A typical system for dealing with user requests would be as illustrated in Fig. 7.5.

The alternative adopted at Stage 3 will depend upon the status of the supervisors in the organisation. For this type of work the supervisor is usually in the best position to estimate the output of his particular gang of men. On the other hand, an independent estimate provides a better and more consistent check on efficiency. In some cases, the supervisor may indicate the scope of the work and the materials required and a work-study officer determines the standard hours by reference to elemental data.

The process also assumes that at Stage 4 it is possible to determine accurately the scope of the work, whereas sometimes, preliminary exploratory work must be carried out before this can be done. In such cases, the job descriptions tend to be very loose so as to cover every possible contingency and as a result are useless for cost control and future estimating purposes. It is important, therefore, that the detailed job description should be entered on the work order as soon as it is known and the estimated labour and material requirements altered accordingly.

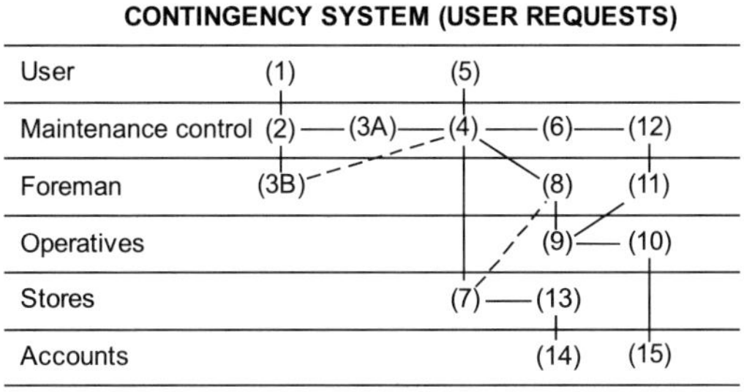

CONTINGENCY SYSTEM (USER REQUESTS)

Fig. 7.5   Contingency system for user requests. (1) User issues request to maintenance control centre describing defect; (2) Request examined by maintenance control to check whether necessary and, if so, priority; (3A) Preview of defect by building supervisor to determine scope of work and job method, or; (3B) Alternatively, preview by trade foreman who will be responsible for supervising execution; (4) Preparation of work order entering date, location, number, job description and estimated labour hours and materials; (5) Copy of work order to user department for information; (6) Copy of work order filed with open jobs; (7) Copy of work order to stores to check availability of materials or to order out-of-stock materials; (8) Copy of work order to foreman to arrange with stores for delivery of materials to site at the programmed time; (9) Instructions to operatives on method of execution of work; (10) Operatives complete time sheets giving details of hours worked, overtime and non-productive time; (11) Foreman enters times for individual jobs on work orders and checks weekly total against time sheet; (12) Work orders for completed jobs extracted from file. (At the same time, progress on completed jobs checked and jobs not started investigated); (13) Stores control procedures; (14, 15) Payroll and accounting procedures.

It is also apparent that where there is a degree of uncertainty a fair amount of discretion must be given to the supervisor to arrange short-term programmes according to actual conditions prevailing at the time. Rigid adherence to programmed dates may not only prove impossible but, if attempted, lead to inadequate work.

Figure 7.6 shows in diagrammatic form an integrated system of work-input control and the related procedures.

7.6 Stock control

The extent to which materials are purchased and stored in advance of use will depend upon the needs of the maintenance organisation, the ease with which materials can be purchased from local merchants and the overall cost consequences. The objects of stockholding are, firstly, to enable materials to be pur-

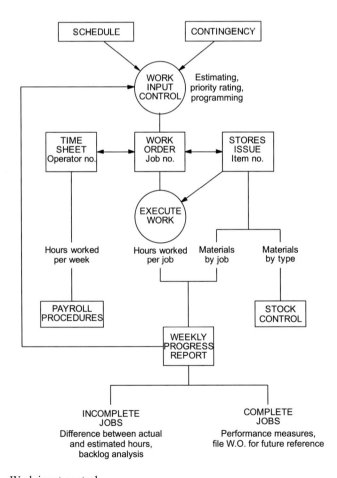

Fig. 7.6   Work input control.

chased in economic quantities and, secondly, to ensure that materials are available as and when required for the execution of urgent repairs.

There are certain advantages to be gained from buying materials in large quantities, such as lower costs for ordering, handling and transportation and perhaps a quantities discount. On the other hand there are costs associated with the holding of stocks, including the use and upkeep of a storeroom, attendant labour, risk of deterioration and obsolescence of the stock, insurance and, by no means least, the interest on capital tied up in stock. It is therefore a matter of determining the 'economic batch quantity' that will minimise the total costs as illustrated in Fig. 7.7.

The economic quantity for ordering may be calculated using the following equation:

$$Q = (2cd/sp)^{1/2}$$

Where:
Q = economic batch quantity
p = item cost
d = annual demand for item
s = stockholding cost expressed as a fraction of stock value
c = delivery cost per batch

For example, assuming the annual demand is 200 items, the cost price per item £5, the stockholding cost assessed as 20% of the stock value and delivery charge £10 per batch; the economic batch quantity would be:

$$Q = \left(\frac{2 \times 10 \times 200}{0.2 \times 5}\right)^{1/2}$$
$$= 63.25; \text{ say, } 65 \text{ items}$$

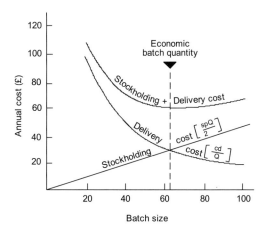

Fig. 7.7   Calculation of economic batch quantity.

However, this assumes that the only variables are delivery and stockholding costs. If there is a discount for quantity it becomes necessary to compare the saving resulting from the quantity discount with the extra cost of holding the additional stock.

Where the rate at which items are withdrawn from the store is constant and replenishments can be made immediately, i.e., the 'lead time' is zero, there is no need to hold a buffer stock for contingencies and replacements should be ordered immediately the stock falls to zero. Usually, however, there is a timelag between the placing of a replenishment order and the receipt of the items in stock. Provided the timelag is predictable it is possible to calculate the quantity which will be used during this period and place the order sufficiently far in advance to ensure that there is always stock available to meet demand. Where, however, the withdrawal rate fluctuates there is a risk that stocks will be exhausted before replenishments are received. This is called a 'stock out' and may result in losses as a result of delay in carrying out urgent repairs. In such circumstances the economic time for placing a replenishment order will be that which achieves a correct balance between the cost of carrying additional safety stock and the cost of running out of stock, as shown in Fig. 7.8.

The simplest system of stock control is that known as the 'two bin' system. This derives its name from the original method of organising the system by physically separating the stock into two bins. Stock was drawn from the first bin until empty and then replenishment stock ordered. When the replenishment items were received the second bin was topped up and the remainder of the items placed in the first bin from which items were issued. This is also known as the reorder level system and may be administered by noting receipts and withdrawals in an inventory, by hand posting or by machine or electronic computer, and ordering replacements when the stock falls below a predetermined level. An

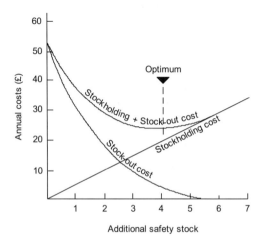

Fig. 7.8   Calculation of optimum stock replenishment levels.

alternative system known as the 'periodic review system' may be used where a series of items are obtained from a single supplier and it is desired to gain the maximum quantity discount. This involves placing orders at regular intervals for quantities calculated to bring the total stock up to some predetermined level.

All withdrawals should be on production of a properly authorised requisition, which should state clearly the description of the material, the quantity required and the work-order number. This can then be used for the purposes of cost accounting and stock control. The only exceptions would be items too small to warrant being controlled in this way, i.e., low-cost items used in large quantities (nails, screws, glass-paper, etc.), which could be issued on demand and charged as a percentage on the value of work done.

A physical check should be made of the items in stock at periodic intervals in order to detect errors in the inventory records. If there are serious unexplained shortages an investigation should be carried out and corrective action taken. At the same time the stock should be examined for obsolescence and deterioration and if necessary disposed of. The risk of this happening will be lessened if steps are taken to achieve the maximum degree of standardisation of components and fittings. Wherever possible replacements should be with standard parts even though some additional minor work is necessary. In the case of proposed new buildings the brief given to the designer should point out the advantages of reducing the variety of finishings and fittings in the context of the overall maintenance policy of the organisation.

References

(1) Handy, C.B. *Understanding Organizations.* Penguin, Harmondsworth.
(2) Wiener, N. (1948) *Cybernetics.* MIT Press, Massachusetts.
(3) Lawrence, P.R. & Lorsch, J.W. (1967) *Organization and Environment.* Harvard University Press.
(4) Ministry of Works (1962) *Survey of Problems before the Construction Industry.* Report by Sir Harold Emmerson. HMSO, London.
(5) Construction Industry Training Board (1975) *Management/Technician Roles in the Construction Industry.* CITB, London.
(6) Layton, E. (1961) *Building by Local Authorities.* Report of an inquiry by the Royal Institute of Public Administration. Allen & Unwin, London.
(7) Department of Health and Social Security (1964) Maintenance of Buildings, Plant and Equipment. *Hospital Technical Memorandum 12.* HMSO, London.
(8) Park, A. (1994) *Facilities Management: An Explanation.* Macmillan Press, Basingstoke.

Chapter 8
Maintenance Procurement

8.1 The procurement process

Maintenance procurement is the process by which required maintenance works are carried out. The procurement process is concerned with the *form* of procurement, whether by contract or direct labour, and with the *quality* of delivery of both the work carried out and the level of service provided. In maintenance works the interaction between the form of procurement and the quality of delivery is complex. Whereas in newbuild, the procurement process is essentially a matter of supplying products and labour to provide a finished building over a certain timescale, in maintenance the process also involves liasing with and working around building users, in circumstances in which the works cannot be specified with certainty. This often may be in the form of small packages of work which, because of the nature of maintenance, may require variations from or clarifications on the original order as a result of discovered circumstances on site during the repair work. For example, work to replumb defective pipework to a sink may escalate if on opening up the work for repair, it is discovered that the defect had triggered rot in surrounding timber work. A further complication lies in the small-scale and multitrade nature of maintenance works that may not fall into conventional building trades divisions. For example the sink repair referred to above may properly require five or six different skills or trades: plumber, joiner, carpenter, plasterer, tiler, electrician – but it would obviously be too costly and impractical to send six different operatives to one small job. Ensuring effective procurement is therefore one of the greatest challenges faced by the maintenance manager.

Although the form of procurement and the quality of delivery are closely bound together, they will be dealt with here in separate chapters. In this chapter, the different forms of contract and direct labour management will be examined. Chapter 9 deals with an analysis of the quality of service delivery, focusing on labour skills and on service delivery in terms of health and safety, timeliness, impact on building users/clients, and monitoring of works done.

8.2 Form of procurement

Maintenance works may range from large planned maintenance projects, for example renewing windows or rewiring electrical installations, to routine annual servicing of boilers, to very small but essential one-off works such as repairing a lock on a front door. Consequently there may be a number of different forms of procurement needed in any one organisation to meet this range of service requirements.

There are a number of procurement routes available for maintenance works:

(1) lump sum contract
(2) measured term contract
(3) cost reimbursement contract
(4) service level agreement
(5) term contracts
(6) direct labour.

The essential feature of the first five of these is the contractual division between client and provider. Even when direct labour is used, this division may operate, for example as required in the delivery of maintenance for local government estates, where compulsory competitive tendering (CCT), introduced during the 1980s, now requires the in-house direct labour organisations to compete on a contract basis with other potential providers from the private sector. Figure 8.1[1] illustrates the procurement routes for day-to-day maintenance taken by six different types of estate.

It will be noted from Fig. 8.1 that there is no consistency of procurement pattern between the estates, which reflects their organisations' different natures and requirements. Measured term contracts (MTC) appear to be favoured by estates with a wide number of dispersed units, such as the housing providers, whereas more concentrated estates on a single site or small number of sites rely more on a dedicated direct labour force. Private estate managers show a preference for lump sum contracts because of the greater ease of accounting and recharging these fixed costs, determined in advance of the works being done. Day works are used mostly by the private estate managers for works that cannot be procured by an advance lump sum contract, notably residential managing agents. It will be noted that the local authorities procure by measured term contract even when their own direct labour is used, and such measured term contracts with their DLOs or with other contractors account for nearly all of their day-to-day maintenance procurement.

For planned maintenance work (not included in Fig. 8.1) lump sum contracts predominate, with a small amount of works procured by the use of measured term contracts.

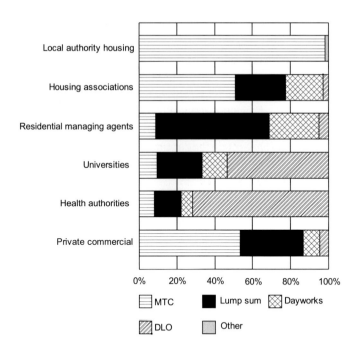

Fig. 8.1   Procurement route: day-to-day maintenance.

8.3 Formation of contract

A contract for building or maintenance work is an agreement between two parties – the builder or contractor and the building owner or employer. The basis of the contract is the offer by the contractor to carry out the work for a certain sum of money and the acceptance of that offer by the employer. It is pertinent to note that the interests of the employer and the contractor differ and are in a sense incompatible. The contractor quite naturally seeks to obtain the contract at the highest price so as to maximise his profits, while the employer is anxious to keep his costs to a minimum consistent with an appropriate standard of workmanship. Clearly there is no advantage to the employer in securing a low price if the result is poor quality work. The essence of the agreement should be the payment of a reasonable sum for work that is properly executed and completed on time.

The offer is usually the contractor's tender and in the case of *Crowshaw* v. *Pritchard* (1899) it was held to be equally binding if described as an 'estimate'. The acceptance by the employer must be made within a reasonable time and must be unconditional. If it suggests new terms it does not create a binding contract. Although there is no legal necessity for building contracts to be in writing, for purposes of record and to avoid disputes it is highly desirable that they should be so evidenced.

It sometimes happens that after a contract has been formed one of the parties discovers that it does not correctly represent his intentions. The most usual type of mistake is the discovery by the contractor of an arithmetical error in his tender build-up resulting in a disadvantageous price. In general, unilateral mistakes of this type do not entitle the party adversely affected to rectification. But where the parties are under a common misapprehension it is open to them to correct their mistake by mutual agreement. In construing a contract the court attempts to ascertain the meaning of the actual words used in the contract documents. Other documents that do not form part of the contract are inadmissible except in so far as they explain the circumstances surrounding the contract or throw light on any special meaning attached to the words used in the contract documents. Thus where quantities are supplied to assist the contractor in arriving at his tender price but do not subsequently form part of the contract, the builder will usually be unable to recover for indispensably necessary work under-measured in the quantities. However, if the quantities are incorporated in the contract they become part of the description of the work and the value of any work omitted from or understated in the quantities is recoverable.

In the vast majority of cases contracts are discharged by due performance. However, a breach of a fundamental term by one party gives the other party the right to treat the breach as a repudiation and sue for damages. For a breach to constitute repudiation it must go to the root of the contract and this would include such actions as complete abandonment of the whole works before substantial completion by the contractor or failure by the employer to give the contractor possession of the site. In normal circumstances breaches such as delayed completion do not amount to repudiation although they may entitle the employer to damages, either liquidated damages agreed at the time of entering into the contract or unliquidated damages awarded by the court. The measure of damages would be losses that arise naturally from the breach or those of which the contractor had actual or imputed knowledge. Where the work has been completed substantially in accordance with the contract requirements, the damages would be the cost of rectification or the difference between the value of the work done and the value of the work which should have been done. Thus in *Hoenig* v. *Isaacs* (1952), in a contract for the decoration of a flat and provision of certain fittings, the employer withheld payment on the ground that parts of the work were defective. The court held that there had been substantial compliance with the contract and that the builder was entitled to the contract sum less the cost of remedying the defects.

Other reasons that a contractor can advance for failure to complete the work include:

(1) *Impossibility.* If circumstances become so radically changed that without default of either party the contractual obligations become incapable of being performed for reasons which could not have been foreseen by the parties at the time of entering into the contract, both parties are automatically discharged. This is a rare occurrence on building contracts and would

not normally cover such events as bad weather, inability to obtain adequate supplies of materials or strikes.

(2) *Illegality.* If at the time of entering into a contract the performance is illegal or contrary to sound morals the contract is void. If its completion becomes illegal after execution of the contract the contract is frustrated and both parties are discharged from further performance.

(3) *Default of other party.* As stated earlier if either party so acts or so expresses himself as to show that he does not mean to accept the obligations of a contract any further the other party may treat it as a repudiation of the contract and sue for damages. Examples of such acts are:

- On the part of the employer – failure to give possession of the site, failure to make proper payments and wrongful dismissal of the contractor.
- On the part of the contractor – complete abandonment of the work before substantial completion.

(4) *Waiver.* This occurs where the parties either expressly or by implication make a new agreement in place of the original contract, either before the work commences or during its progress.

In addition to the contractual liabilities it is also necessary to consider possible liabilities in tort. A tort is a civil wrong that occurs by reason of a breach of a general duty that is owed to society as a whole. It is therefore much broader than a contractual obligation, which is owed only to the other party to the contract. The more common heads of liability in tort include negligence, nuisance and trespass.

(1) *Negligence.* A professional person must apply a fair, reasonable and competent degree of skill and care in the exercise of his duties and the usual way of testing this is whether or not other persons of the same profession would or would not have done the same thing in the circumstances. This duty is owed to everyone who might be affected by the wrongful act. Recent judicial decisions have resulted in an extension of the time during which persons concerned with the design, construction or inspection of buildings may be liable. The Limitations Act 1980 provides for a period of limitation of six years for simple contracts and twelve years for contracts by deed from the date of the cause of action. The date of the cause of action was originally taken as the date of completion of the contract but in the case of *Pirelli General Cable Works Ltd* v. *Oscar Faber & Partners* (1983), it was held that a cause of action in tort for negligence in the design or workmanship of a building accrued at the date when physical damage occurred to the building whether or not that damage could have been discovered at that date by the plaintiff. The uncertainty as to the total period of liability created by such judgements has been resolved by the Latent Damage Act 1986 which provides that legal action will be allowed over defects up to three years after they are discovered, or should have been discovered, with a

'long stop' rule which will bar actions more than fifteen years after the alleged bad work.

(2) *Nuisance.* Building operations are often noisy and dirty and likely to give offence to adjoining owners. However, provided all reasonable steps are taken to minimise the inconvenience the neighbours must put up with it.

(3) *Trespass.* Except for the many statutory organisations that have powers of entry no one is entitled without leave to cross the boundaries of another person's land. This includes placing or allowing materials to get onto another's land and the remedy for so doing is damages and/or an injunction to secure discontinuance of the act.

Generally one is not responsible for the torts of an independent contractor but there are exceptions to this rule. Where the work is by its nature dangerous and likely to cause damage to other property, employers have a strict liability in the event of damage being caused and they cannot escape this liability by employing someone else to do the work. This applies particularly to work to party walls or where hazardous techniques are being used. In such cases it is important to ensure that the contractor has taken out insurance in the joint names of both the employer and the contractor.

8.4 Types of contract

The main differences between the various types of contract lie in the methods of evaluating the work and the degree of financial risk borne by the contractor and the client respectively.

Lump sum contracts

The contractor agrees to execute the work for an agreed sum based on information derived from drawings, specifications, bills of quantities or site inspection. Lump sum contracts are used for large, predetermined works such as electrical rewiring, reroofing, or window replacements, commonly as part of a planned maintenance programme. The method presupposes that there is sufficient information on which to assess accurately the scope of the work involved. Small areas of uncertainty may be left to the discretion of the contractor to assess the risk and price accordingly or may be covered by the inclusion of a provisional sum. In the latter case, the cost of the work would be ascertained after execution on the basis of day work, or in accordance with a schedule of rates submitted with the tender or by agreeing a reasonable sum. However, while isolated provisional sums are admissible for minor items, a disproportionate number of such sums would completely destroy the primary advantage of this type of contract, i.e., knowledge by the client of his financial commitment before the work is commenced.

Measured term contracts

This type is also called a measure and value contract and is useful where details are too scanty to permit the preparation of a precise specification at the time of commencing the work. It is widely used in the procurement not only of day-to-day repairs, but also for larger works such as refits to void properties. The schedule lists all the items of labour and material which are expected to be required and may be an *ad hoc* schedule for a particular job based on past experience of similar jobs or a standard schedule designed to cover a wide range of jobs. The schedule may be unpriced, in which case the contractors tender by inserting a rate against each item, or there may be standard rates included in the schedule and the contractors tender by quoting a percentage on or off the standard rates, such as the national schedule of rates[2] used by many local authorities. The work is measured on completion in terms of the schedule items and priced out either at the contractor's rates or at the standard rates plus or minus the quoted percentage. The cost is, therefore, not known until the work has been completed although the approximate quantities of the various items may be estimated prior to commencement for cost control purposes.

The National Schedule of Rates[2] is an example of a standard prepriced schedule and it has been suggested that it could be used with advantage by private firms. Although a convenient way of providing in advance for the pricing of work of uncertain scope it lacks estimating accuracy. The contractors tendering are required to state their offers in the form of one or two percentages on or off the schedule rates as a whole. The presumption is that an overall percentage adjustment will bring the schedule rates into line with those normally charged by the contractors. This is unlikely to be the case for the following reasons:

(1) There is little uniformity among contractors regarding the pricing of individual items of work and inevitably some of the schedule rates will be higher and some lower than a particular contractor would normally charge.

(2) The mix of items for a particular job may well exaggerate the effect of differences in pricing patterns. For example, a job may consist largely of those items for which the standard rates are higher than the contractor's usual rates and the addition of a percentage would serve only to make the discrepancy even greater.

(3) The final cost will depend upon both the quantity and the rate for each item comprised in the work executed. If one accepts that some of the schedule rates will be 'high' and some 'low' then the relative quantities of each will affect the outcome.

(4) The schedule rates are essentially averages and do not reflect the particular conditions under which individual items of work will be executed.

(5) Standard rates tend to be updated at infrequent intervals and the pattern of pricing becomes increasingly inappropriate because of differential changes in the prices of materials and new working methods.

In order to arrive at a realistic estimate, the contractor tendering should therefore estimate the likely proportions of the schedule items in the work which is likely to be ordered and then determine a percentage which will equate the cost based on the schedule rates to the cost obtained using his normal rates. Clearly this process is shrouded with uncertainty and it is probable that a contractor would merely identify the predominant trades and base his percentage on the estimated quantities of the major items in these trades.

However, in spite of the above defects, a schedule fulfils a number of useful functions. Principally it establishes standard work descriptions and lays down a pattern of pricing which, although not corresponding precisely to the pattern of prices normally adopted by the individual contractors, is sufficiently close to form an acceptable basis for valuing work executed. It also removes a considerable workload from the contractor in pricing a schedule that may not be accepted, and therefore makes it more likely that contractors invited to tender on the basis of the schedule will in fact do so. The percentage additions also provide a ready means of establishing trends in maintenance prices. Additionally, the schedule is automatically updated in line with inflation, enabling keener prices to be submitted by the contractor in the knowledge that the risks of being committed to an underpriced schedule in case of labour or materials price inflation during the term of the contract is limited.

Cost reimbursement contracts

The contractor carries out the work and is paid the prime cost of labour and material plus either a previously agreed percentage or a fixed fee to cover overhead charges and profit. The obvious disadvantage of this type of contract is the absence of any financial incentive to encourage the contractor to carry out the work as economically as possible. Its use should therefore be restricted to small or extremely urgent jobs for which no other method is possible. However, there may be occasions on which the nature of the work is so uncertain that, in an effort to provide cover against every eventuality, a firm price might be in excess of that obtained on a cost-plus basis. Also, although incentive may be lacking on an individual job the position would be different if there were the prospect of a series of jobs. Certainly this type of contract demands the employment of a reputable contractor and close scrutiny of the contractor's account to ensure that the labour and materials charged are reasonable in relation to the work done.

The additional cost of supervision is another factor that should not be overlooked. For lump sum contracts the supervisory function relates mainly to quality and checking that the work is being done according to instructions. For cost reimbursement contracts the supervisor's duties would extend to making sure that the work is being done in the most economical manner with the minimum wastage of materials. Such interference in the manner of executing the work is justified by the fact that extra costs arising from inappropriate working methods are borne by the client and not by the contractor.

Also, the practice of making the selection solely on the basis of the percentage addition quoted for overheads and profit is questionable. Clearly a contractor who quotes a marginally higher percentage than his nearest competitor might organise the work more efficiently and produce a lower prime cost which, in spite of the higher percentage addition, would result in a lower total cost. Indeed, it is probable that the higher percentage reflects a more sophisticated management structure that would be capable of achieving this desirable end. It is necessary, therefore, to ensure that there is equivalence of management skills before accepting the contractor who quotes the lowest percentage.

The main weakness is the difficulty of checking the correctness of the contractor's account, particularly with regard to the number of labour hours. A method that has proved beneficial is known as 'controlled day work' whereby all jobs are pre-estimated prior to issuing the order. Then if the contractor's account exceeds the estimate by more than a certain percentage the reasons are investigated. The permitted variation may range from 10% for the larger jobs to 20% for the smaller jobs. If the estimates are based on the recorded costs for similar jobs with suitable adjustments for differences in working conditions they give a positive check on the reasonableness of the contractor's charges. However, the fact remains that the major risks are borne by the employer, e.g. the extra cost of overcoming difficulties which impede the progress of the work are recoverable whereas in many cases these can be regarded as legitimate business risks. As this is a form of contract that does not encourage price efficiency, it should be only rarely used by the maintenance manager.

Service level agreements

This is an increasingly used method of procurement, intended to reduce the considerable amount of paperwork involved in administering the other types of contract where the administrative costs can exceed the cost of actually executing the work. The system involves agreeing a lump sum with a contractor for undertaking a range of recurring works of a similar kind to a specified group of buildings over an agreed period. The method is somewhat similar to insurance in that the contractor agrees to carry out all the work of the specified types with no adjustment of the contract sum. It is used for example in maintenance and servicing contracts for domestic heating systems in a scheduled list of properties, whereby the contract specifies one annual service to each boiler, and thereafter call-outs and breakdowns are paid for on a materials only basis; the cost of the labour being borne by the contractor under the lump sum contract. Its use is common in services maintenance such as for lifts, but is now spreading to more general maintenance works.

Depending on the method of estimating, this form of procurement may advantage contractors who are already engaged in the work which is being tendered, because they will be familiar with the detail of the properties' installations and thus will be in a better position to assess the risk of losing profit on call-outs. New entrants to the tendering process would need to carry out extensive investigations to ascertain this risk prior to tendering. The contractor is paid one-twelfth of the annual lump sum each month although inspections are made from time to time to ensure that the work is being completed satisfactorily.

The advantages claimed for this system are:

(1) There is a considerable saving in administrative time with the result that qualified personnel are released for other duties.
(2) Payment to contractors is made promptly each month without retention and without the need for detailed measurements or checking accounts.
(3) It is thought that there is some improvement in the quality of the work since it is in the contractor's interest to reduce the probability of expensive failures by early preventive maintenance.
(4) It improves tenant satisfaction in that repairs are executed promptly and in some cases on the initiative of the contractor.
(5) There is a direct incentive for the contractor to do the work as economically as possible.

However, it is probable that unless the number of units to be maintained is very large there will be significant deviations from year to year in the total amount of work required. There may also be differences of opinion as to whether a particular item of work is necessary. This could happen in those cases where a component deteriorates at a slow rate and has reached a point where, while still functionally satisfactory, it is considered by the user to be below an acceptable standard. Such disputes are likely to become more prevalent towards the end of the specified period. There is therefore a need to specify closely the required performance levels and criteria for judging whether these have been met for the items included in the service level agreement as a precursor of the tender.

Clearly the successful operation of such a contract calls for the utmost goodwill and understanding on both sides. The obligations of the contractor are open-ended in that in return for a fixed annual sum he or she agrees to carry out either on notice from the superintending officer or on his or her own initiative all work of very broad and ill-defined types, e.g. all cleaning of gutters and gullies, repair, renewal and maintenance work to the fabric, fittings, finishings, hot and cold water and heating services, and drains. There are certain exceptions, principally decorations where not associated with an item of repair or renewal, and any single item exceeding an agreed value.

Term contracts

Under this type of contract the contractor is given the opportunity to carry out all work of a certain type or falling within certain limits of cost for an agreed period.

The work done is usually priced on either a schedule (measured term) or a cost-reimbursement (day work term) basis, although for the larger jobs it may be more advantageous to negotiate a lump sum.

The main advantages claimed for term contracts are:

- lower prices quoted by contractors in consideration of the benefits of an assured programme of work. This is probably true, but the estimating errors are likely to exceed any marginal allowance that might be made on these grounds. Where a schedule is to be used for the valuation of work executed, the quoted percentage addition must allow for fluctuations in wage rates and materials prices over the contract period (usually two years). As wage rates and individual material prices will vary at different rates and their proportions in the ultimate workload are not known, the probability that the percentage represents the correct amount is clearly very low.
- saving in time and overheads compared with that entailed in arranging single-job contracts. However, although there would obviously be some reduction in precontract time, the time required for the measurement of completed work for a schedule contract or for supervision of and checking accounts for cost-reimbursement contracts would be little affected. Where the work can be specified with reasonable precision a negotiated lump sum would probably give the greatest total saving in time.
- the long-term relationship results in the contractor becoming increasingly familiar with the building and the needs of the occupants. In this it confers some of the benefits of a directly employed labour force, particularly if the same operatives are employed throughout the term.

The primary disadvantages are:

- the reasonableness of the contractor's rates are tested in open competition only at infrequent intervals and may become increasingly unfavourable.
- if a contractor's employment is terminated for any reason, a number of partly finished jobs will be left which will have to be completed by another contractor at an increased cost.
- towards the end of the contract period the number of orders, especially for the larger jobs, will tail off and at the beginning of the next period will reach their normal level only when the new contractor has settled in.

However, on balance it would seem that the term contract has considerable advantages in the context of a planned maintenance system. The special relationship subsisting over a period of years permits the contractor to participate in the planning process and possibly suggest alternative timings for the work, resulting in some cost savings.

Usually contractors who have been previously vetted for inclusion in an approved list would be invited to tender. The essential information that should be given to contractors tendering for term contracts is:

(1) Form of tender and instructions relating to date and mode of submission and form of acceptance.
(2) General conditions of contract to be used. Where these have been specially drawn up a copy should be supplied to the tenderers.
(3) The contract area and details of the buildings included in the contract.
(4) The contract period and commencement date. Usually the term would be from two to three years with provision for the annual updating of the rates on an agreed basis, e.g. by reference to a published index.
(5) The likely total annual value of work that will be ordered. Although this is not legally binding it should be borne in mind that the contractor will have taken this into account when fixing his percentage addition to the schedule rates.
(6) The maximum value of any single order above which a lump sum quotation may be required.
(7) The method of valuation. The primary means of valuation will be by reference to the schedule of rates on which the contractor will be required to quote a percentage adjustment. The aim should be to cover 80–90% of the jobs with the scheduled rates, with the day work element restricted to those jobs which are of such an uncertain nature or executed under such diverse conditions that a schedule rate would be inappropriate.
(8) Specification of materials and workmanship. This would follow the usual pattern of good building specifications but covering items of unusual character particular to this type of contract, e.g. constraints imposed by working in occupied buildings.
(9) Schedule of rates. This may be a standard prepriced schedule or one devised to meet the particular requirements of the building owner. In the latter case a schedule based largely on numbered composite items can reduce the administrative costs of measurement and valuation.

Direct labour

Large direct labour organisations (DLOs) were frequently used by local authorities and large corporations to procure maintenance works. However, over the last twenty years, restructuring in such organisations has resulted in the direct labour operations either being scaled down dramatically, or set up as separate business entities in their own right, either as private companies or still within the authority. The chief driving force behind this has been CCT, which has obliged local authorities to set up a client/contractor interface between the council as provider and the DLO as one of many potential contractors[3]. There is little doubt that compulsory competition has forced many DLOs to look at the cost effectiveness of their service delivery and to review working practices and bonus payment schemes in the light of equivalents in the private sector. However the orthodoxy of assuming that competition *per se* provides a better level of service has not always been justified in practice. A formal contract may limit the amount of flexibility in the delivery of a maintenance service that the contractor is prepared to offer, which may lead to gaps in provision, particularly in relation to emergency works as a result of unexpected freezes or flooding, for example; where DLOs would be able to provide a focused and dedicated response to such events. Additionally, the goals of the DLO may be more directed towards providing a good service than in making a profit in providing just the minimum level of service specified in a contract, which can lead to cost and corner-cutting in order to maintain profit margins. This has subsequently been recognised in some local authorities which have gone wholly down the competition route and who have found that the cheapest tender does not necessarily give the best value in terms of service provision and user satisfaction. In this case the strict divide between the client and contractor has been softened by the concept of partnerships and 'best value' philosophy whereby the quality of service is paramount, and the price is secondary to this aim.

Some organisations both in the public and private sectors therefore have directly employed operatives within the maintenance management organisation to fulfil certain service level functions. This may range in size from large numbers of fitters, electricians, etc. in manufacturing industries down to individual caretakers/handymen employed to service a single block or group of buildings.

The advantages of having such a direct labour force are:

- they are employed by the client organisation; therefore their goals coincide with those of the client organisation
- they develop close working relationships with the users of the buildings
- they become familiar with the specific characteristics of the buildings over time
- they are on first call in the case of emergency rather than having a number of clients competing for their services.

Set against this are certain disadvantages of employing operatives directly:

- without the discipline of free-market price competition they may become economically inefficient in their delivery
- there needs to be a basic minimum workload to ensure they are in continuous work, as idle time will still have to be paid for by the client maintenance organisation. In this respect it is more usual for workers in small DLOs to be general handymen, rather than to have a single skill or trade for which the demand in the estate may be sporadic
- without a large infrastructure there may be little opportunity for career advancement and therefore little incentive to reskill or to work efficiently

- there will be an ongoing base cost for the provision and maintenance of plant, equipment and vehicles whether they are in use or not.

Where DLOs are separate business entities, either within parent organisations or as companies in their own right, the client maintenance manager will usually be obliged to deal with them as he or she would with any other contractor. Within the maintenance management organisation, a DLO will require careful management particularly in relation to base workload, work efficiency, health and safety, and managing establishment costs such as plant and personnel management.

8.5 Suitability of contractor

The basic problem of selection is to predict which contractor of those available is most likely to achieve the client's objectives. It is necessary, therefore, to compare contractors' known or assumed abilities with the services required of them. This process involves competition between those contractors who wish to secure the contract. The competition may be formal and based on criteria that can be objectively measured, or it may be informal and based on subjective judgement. Even where a building owner approaches a single contractor there will be some degree of implied competition in that the mere choosing of a particular contractor suggests that there are reasons for rejecting all the other contractors who are equally capable of undertaking the work.

The contractor's suitability should be considered under three criteria:

- Does the contractor have the potential resources necessary for the performance of the services required?
- Is the contractor likely to apply these resources adequately to the contract and are there any reasons why he or she should not do so?
- What are the contractor's specific proposals for the contract and are they reasonable?

To determine the degree to which a contractor meets the above criteria, it is necessary to consider the following aspects of his organisation. The contractor should have a proved record of successfully performing the type of work required. A survey by the Research Institute for Consumer Affairs[4] revealed that in fact building firms were mostly selected on the basis of either personal knowledge or recommendation and that, by and large, clients were satisfied with the work done.

Financial stability

The ease with which anyone without capital or experience can set up in the building business has been commented on in many official reports. Equally, the very high bankruptcy rate is a matter of public record. It is quite clear that many small firms concerned with maintenance have very little working capital and that bad debts or long delayed payment can affect critically their financial situation. Most firms in this category are private concerns that rely on short-term loans and overdrafts from banks and on their own personal savings for finance. Further assistance is obtained from credit facilities provided by builders' merchants, and their willingness to supply materials is a fairly reliable indicator of the financial standing of the builder. However, in the absence of more detailed information, one would have to form an opinion of the builder's stability on the basis of such factors as length of time in business. In the case of large contracts it is usual to specify the deposition of a performance bond, guaranteed by a major bank or lending institution, which indemnifies the client against potential extra costs and disruption caused if the builder were to be declared bankrupt during the course of the contract.

Resources

The contractor's resources may be considered under the following headings:

(1) Physical – the nature of the premises from which the contractor conducts his or her business, including offices, workshops and stores. Also, the plant and equipment owned by the contractor and his access to building materials.

(2) Human – the operative skills available in the contractor's organisation and their suitability for the work to be undertaken. This is dealt with in more detail below.

(3) Management – the number and qualifications of managerial and supervisory staff and the extent of their technical knowledge and experience. This would be particularly important where the contractor is engaged for a term of years and is required to participate in the planning of the work.

Scope of work

This would involve an assessment of the type and size of jobs normally undertaken by the contractor and the work that he or she usually sublets. Firms can be broadly divided into general builders capable of dealing with all the traditional trades and specialist firms. General builders may be further subdivided into small firms capable of undertaking jobs up to about £500 in value and medium-sized firms for the larger jobs.

There has been a tendency in recent years for firms to specialise in certain aspects of what were hitherto regarded as general building, e.g. drain rodding, remedial treatment of timber, etc. By their greater experience in a limited field of operations they are usually able to offer a better service than a general builder, but this may be offset by their inability to carry out the whole of a particular repair job and the subsequent necessity to employ other contractors in addition.

Availability

There is little advantage to the building owner if the contractor possesses all the necessary resources but is unable because of concurrent commitments to use them when required. It not infrequently happens that small builders accept more work than they can conveniently deal with, to provide a backlog of several weeks' duration and so ensure continuity. This delayed response provides a useful means of smoothing the flow of work to the operatives and in an uncertain market enables the contractor to plan at least in the short term. For certain types of work the response time will vary according to the time of the year and may be very much shorter in winter. At such times the fall in demand which leads to a reduction in the response time also results in lower prices. There is thus a price advantage to be obtained from planning work so that it can be executed during a slack period, provided that it is not detrimental to the use of the building or to the quality of the work done.

Co-operativeness

Although a very difficult quality to assess in the absence of personal experience, it is essential that there should be harmonious relations between the building owner and the contractor. Often the nature of the work is not susceptible to precise definition at the time of selecting the contractor and the actual extent of the work may not become apparent until surface finishings have been removed. There is thus a need for close liaison between the client organisation and the contractor in devising methods that will be both technically and financially satisfactory. Co-operation in maintenance works is beneficial to both parties in view of the long-term and repetitive nature of building maintenance works, which means the client will be in a position to supply regular tender opportunities to contractors whose level of service proves to be good in practice.

Price levels

In many cases price is the sole criterion by which contractors are judged, but it should be noted that if the price is too low it may inhibit the attainment of the building owner's other objectives relating to quality and time. Orchard[5] makes the point that unless care is taken over the selection of the contractor, what appears to be a keen price can lead to a lot of subsequent trouble and expense. The price is made up of the basic cost to the contractor of labour, materials and plant plus an allowance for profit. The basic cost is a reflection of the ability of the contractor to organise the work so as to reduce ineffective time to a minimum, while the profit element will to a large extent indicate his keenness to secure the contract and his workload at the time of tendering. In addition, the price will be affected by the skill of the contractor in estimating the basic cost and the price levels of his competitors.

8.6 Mode of selection

It is perhaps axiomatic to state that selection should be by those criteria that are least costly to apply.

Degree of competition

Tendering procedures may be classified according to the degree of competition as follows.

Open tendering
This is where a job is advertised and all contractors are free to quote without any prior enquiry as to their competence. The method has been widely used by local authorities in that it fulfils the requirement of public accountability and removes any suspicion of favouritism. However, the method is subject to criticism in that the low prices resulting from indiscriminate tendering are reflected in the quality of the work done and that resources are wasted when too many firms tender for the same job.

Selective tendering
In this method the client invites tenders from a limited number of reputable firms. The list of contractors may be drawn up specifically for a particular contract or the client may have a standing list of approved contractors from which a short list is drawn up for each contract. Separate lists may be kept according to the type of work or size of jobs that the firms normally undertake. Where a standing list is used it should be reviewed at intervals so that progressive new firms are not excluded and also so that unsatisfactory firms can be removed from the list.

Negotiation
This is where the client invites a tender from a single contractor who is known to have the necessary qualities, abilities and resources to carry out the work satisfactorily. Although there is no competitive element present and as a result the prices may be somewhat higher than with the other methods of tendering, it permits a greater involvement of the contractor during the planning process and perhaps speedier completion and better quality work.

The number of stages

Broadly, methods may be single-stage, in which a single contractor is approached, or multistage, in which the characteristics of a number of contractors are compared in turn with criteria representing the building owner's requirements until eventually only one contractor remains as satisfying all the criteria.

The type of criteria used

This will vary from the single criterion of price to a combination of subjective assessments of the various characteristics of the contractors under consideration. The characteristics may be assessed on a numerical scale and each weighted according to its importance to the building owner. The weighted individual factors may then be totalled to give a single unit for the purpose of comparing the overall suitability of the different contractors.

The stage at which the contractor's offer is made

This will depend upon whether or not it is possible to specify fully the work before inviting tenders and starting on site. Where this is possible and fixed-price tenders are invited, the selection of the contractor and the acceptance of the offer occur simultaneously when a particular tender is accepted. Where the degree of uncertainty is such as to preclude this or where it is desirable that the contractor should give preliminary advice on the scope of the work, the selection of the contractor and the submission of an offer by that contractor will be separated in time. In this latter case, if price is to be a criterion it will be necessary when making the preliminary selection of a contractor to examine the methods that will subsequently be used for building up prices. Thus in addition to general information on the contractor's resources and reputation more specific information is required on percentage additions for overheads and profit.

The nature of the offer

This will be determined by the type of contract that the building owner has elected to adopt. The offer may be for a fixed price, either stated as a lump sum at the time of tendering or to be arrived at on the basis of a schedule of rates after completion, or it may be for an indeterminate amount arrived at on the basis of the actual cost of labour and materials used plus a percentage or fixed fee for establishment charges and profit. The offer may relate to a single job or to a series of jobs of a certain type and within a specified price-range over a period of years. However, as emphasised by McCanlis[6], 'Any client who has a long term programme of work would be well advised in the interests of economy to plan his programme to achieve a steady flow of work commensurate with the resources of the selected contractors.' Where a building owner does not have a sufficient quantity of work to make this possible he or she is dependent upon the contractor's receiving work from other clients to provide the necessary continuity.

Implied terms

For small isolated jobs the contract may be verbal or contained in letters passing between the parties. In the absence of express conditions, the terms implied at common law are:

(1) The building owner must allow the contractor to enter the building at the necessary time for the purpose of executing the work; he or she must give necessary instructions within a reasonable time and not obstruct the contractor in the performance of the work, and must pay a reasonable price.
(2) The contractor must do the work in a professional manner and complete within a reasonable time.

In addition, where the contractor supplies materials there is an implied warranty that the materials are:

(1) reasonably fit for the purposes for which they will be used, and
(2) of good quality.

These common law obligations have been codified in the Sale of Goods Act 1979, which implies a condition that goods shall be reasonably fit for any purpose made known to the seller by the buyer, and the Supply of Goods and Services Act 1982, which provides that, in the absence of express provision, anyone providing a service shall carry out the service with reasonable skill and care, within a reasonable time and at a reasonable price.

However, if the building owner does not rely upon the contractor's skill and judgement in choosing the materials the fitness warranty is excluded. This principle was laid down in *Young and Marten* v. *McManus Childs Ltd* (1968), in which the builder had been ordered to tile houses with 'Somerset 13' tiles which were apparently perfect on delivery but which later developed weathering defects necessitating their replacement. It was held that the builder would not have been responsible if tiles of good quality had been unsuitable for their purpose because the employer had chosen them, but he was liable for the defect in quality even though it was a latent defect undiscoverable by proper care on his part. In *Gloucestershire County Council* v. *Richardson* (1968) it was held that where the contractor is required by the building owner to obtain materials from a supplier on terms which severely restrict his right of action against the supplier, the quality warranty may also be excluded. The reason for this decision is that normally where the contractor is required to reimburse the employer in respect of defective materials he or she can recover the loss from the supplier who in turn may have a remedy against the manufacturer. However, where there is a collateral warranty the employer may sue the supplier direct. Thus in *Shanklin Pier Ltd* v. *Detel Products* (1951) the defendants warranted the suitability of their paint for repainting a pier in consideration of the plaintiffs' instructing the contractor to use the paint. In the event the paint proved a failure and the pier company recovered damages for breach of the warranty.

For contracts of any size it is highly desirable that the terms should be in writing and expressly stated to avoid any controversy at a later date over what was intended.

8.7 Usual conditions

The object should be to provide a fair and equitable legal framework which will ensure that the work is carried out in a proper manner and that the contractor will receive a reasonable fee for his efforts. It is not intended to deal in detail with the legal relationships between the building owner and the contractor, but merely to point out those facets that have particular relevance in the context of planned maintenance. These are the related aspects of scope of work, price, time, quality, risk and disputes.

8.8 Scope of work

It is an essential requirement of any contract that there should be a clear understanding by both parties of their respective obligations. Thus the work which the contractor will be required to undertake should be as clearly defined as the circumstances permit. However, in some instances, full details will not be available at the time when the contractor is selected. This will obviously be the case where the contractor makes a standing offer to carry out work as and when ordered for a specified period. In such circumstances, while past experience will suggest the probable total workload for the period, the timing and extent of individual jobs will be uncertain. It will thus be necessary to agree a formula for arriving at a price for the work when details are available.

Usually the larger jobs will have been subjected to close scrutiny when preparing the budget estimates and the actual scope of the work determined by inspection. Where competitive tenders are to be invited the work will be programmed so as to allow sufficient time for the preparation of contract documents. However, there is no logical and systematic approach to the problems encountered in describing maintenance work. Methods vary from a detailed schedule of the individual items comprised in the work with the quantities of each item stated, to a broad general statement of the end result to be achieved.

The practical effect of these different methods lies in the amount of discretion allowed to the contractor. Where the work is stated in broad terms the onus is placed on the contractor to decide the precise extent of the work necessary and he or she is contractually liable to do all work indispensably necessary to achieve the end result. The detailed approach removes this discretion and, if the schedule is a contract document, the contractor is entitled to extra payment for work not specifically included therein. Clearly, where the work is described in broad terms variations in tenders will reflect not only differences in price levels but also different interpretations as to the amount of work involved. Where there is a large area of uncertainty there will be a natural tendency to inflate prices so as to cover every possible eventuality.

Where detailed information is given, it takes the form of separate statements about different aspects of the work which must be pieced together and usually supplemented by a site visit in order to obtain the complete picture.

The statements relate to:

(1) Location
   - drawings
   - location references in the specification
   - subdivision of the specification or the schedule of items according to location
(2) Work
   - description of operations
   - extent of operations
   - standard of workmanship
   - restrictions – method/time
(3) Materials
   - type
   - tests

Much of the above is susceptible to standardisation; in particular it would be advantageous to standardise work descriptions and clauses relating to workmanship and materials. However, the traditional specification is too loose a form of communication and lacks the consistency necessary for this purpose. Attempts to speed up the process of producing specifications by standardising clauses have usually taken the form of comprehensive spot items that are incapable of adjustment to suit different circumstances.

A major concern of the building owner is that the price paid is fair and reasonable in relation to the amount and quality of work done and is within the budget estimates. The application of continuous cost control requires clear conditions relating to:

(1) A statement of the lump sum or the formula which is to be used to arrive at the contract sum postcompletion. Where the offer is in the form of a lump sum there should be an itemised breakdown of the work with each item separately priced for cost control and feedback purposes.

(2)   A statement of the documentary evidence that must be produced by the contractor as a prior condition to payment where work is postpriced, e.g. submission of certified time sheets and invoices in respect of materials for cost reimbursement contracts.

(3)   The periods at which interim payments will be made and the method by which they will be valued. For small jobs, which can be completed within the space of one month, interim payments are not necessary but for larger jobs there is an implied right to payment from time to time. Usually these payments would be made either at monthly intervals or when the contractor has reached an agreed stage in the work.

(4)   The procedures for ordering variations in the quality and quantity of the work and, in the case of lump sum contracts, the method of valuing such variations. For schedule and cost-reimbursement contracts, the contractor would be reimbursed automatically. A particular problem that might arise is whether or not the work is included by implication. In many cases the descriptions are so loose that it is difficult to determine precisely what was intended. As a general rule if the work can be shown to be indispensably necessary to effect completion, it would be included by implication even though not specifically described.

(5)   In addition, where the contract is of a cost-reimbursement type, there should be clear agreement as to what matters may be claimed as prime cost and what are deemed to be included in the percentage addition. Also it is desirable to give the client organisation greater control over the use of resources than would be appropriate for other types of contract. This control would extend to the method of working and sequence of operations, the number of people employed on the site, the purchase of materials and the plant used on the site.

Time

The timing of the larger items of work will have been fixed when drawing up the master programme, with the object of causing as little interference as possible with user activities and also to ensure that payments for work done phase in with the overall cash flow pattern of the firm. It is probable therefore that failure to complete by the specified time could result in losses to the firm. To cover this contingency, it is usual to provide for the contractor to pay the building owner an agreed amount for each week of delay beyond the date of completion. However, it is important that the amount of liquidated damages should be a genuine pre-estimate of the actual loss likely to be suffered, otherwise it may be construed as a penalty and be unenforceable. In addition, there should be a parallel clause empowering the building owner to extend the time for delay caused by specified events beyond the control of the contractor. Otherwise, the ordering of additional work by the building owner would render the original date of completion inappropriate and, in the absence of provisions for the substitution of an extended date of completion, the building owner would lose his right to claim the damages.

Where the agreement is to carry out work as and when ordered, it is not possible to lay down commencement and completion dates at the time of entering into the contract. Such dates could be given when ordering the individual jobs but reliance would have to be placed on the contractor's obligation to complete within a reasonable time. The building owner may be given a contractual right to determine the contract if he or she considers that progress is not satisfactory, but in practice such a power should be exercised with caution to avoid a claim for wrongful dismissal.

Quality

The contract should contain appropriate safeguards to ensure that the quality of the work done is satisfactory. A primary obligation into which the contractor is normally required to enter is to 'carry out and complete the works to the reasonable satisfaction of the building owner or his or her representative'. What is reasonable would depend on the surrounding circumstances and, in particular, on the price paid.

Conditions which seek to safeguard the employer's interests in this respect are of two types: those which attempt to prescribe the circumstances which will favour the production of a reasonable quality, e.g. a requirement to keep a competent foreman on the site at all times, and those which provide a building owner with a remedy if the quality is unsatisfactory, e.g. contractor to make good any defects caused by failure to comply with the specification for an agreed period following completion of the works.

Additionally, where there is no final and conclusive certificate the contractor's liability for work not in accordance with the contract extends for a period of six years in the case of a simple contract and twelve years if the contract is by deed from the date of the cause of action as provided in the Limitation Act 1980. Where fraud can be imputed, the foregoing periods would run from the date of discovering the fraud.

Risk

The execution of building work inevitably involves some risk of personal injury or damage to property. It is essential that such risks should be adequately covered by insurance, preferably by a joint names policy to cover all third-party liability irrespective of who is negligent. In the case of maintenance, the existing building would normally be insured against damage by fire and other common risks and additional cover could be arranged for the extra risk created by the building operations.

While it is a general rule of law that a person is not responsible for the torts of an independent contractor, there are exceptions that could be important in the context of maintenance.

Thus, the building owner may be liable:

(1) if the work causes loss to a third party and the employer has not imposed a duty of avoiding such loss on the contractor
(2) where the work of its very nature involves a risk of damage to a third party, e.g. interference with the right of support of an adjoining building
(3) where fire is negligently caused on the site by the contractor in the performance of the contract and it spreads, causing damage to adjoining buildings.

These areas of strict liability should be covered by insurance rather than relying on an indemnity from the contractor which in the event may prove valueless.

Another type of risk that may have to be guarded against applies where the operatives may gain information on processes which would be of interest to the firm's competitors, or which may involve matters of national security.

8.9 Disputes

Many maintenance contracts are so loosely worded and the scope of the work so ill defined that the smooth running of the contract calls for a good deal of give-and-take on both sides. It is perhaps only the smallness of the sums involved that dissuades the parties from engaging in costly litigation. For jobs of any size it is desirable to lay down procedures for dealing with any disputes that may arise, usually by providing that they should be referred to a named arbitrator or one who is to be selected in an agreed way. This method of settling disputes is generally regarded as quicker and cheaper than the ordinary processes of the law.

8.10 Contract documents

The contract documents will depend upon the size, nature and complexity of the work and whether payment is to be made on a lump sum, schedule or cost reimbursement basis. They may include a separate form of agreement with attached conditions, drawings, specifications, bills of quantities and schedules of rates.

Agreement and conditions

It is advantageous to use a standard form of agreement wherever possible. Unfortunately there are few standard forms available that cover the types of contractual arrangements required for the general run of maintenance work. The principal standard forms are as follows.

JCT Standard Form of Building Contract

This form is issued by the Joint Contracts Tribunal, which is a committee composed of representatives of the RIBA, the Construction Confederation, RICS, British Property Federation and various local authority and subcontractors' associations. There are three variants of the form (with quantities, without quantities, and with approximate quantities). The form is published in two editions, one for private use and the other for local authority use. In addition the JCT has issued a complementary set of documents for nominated subcontracts – a standard form of tender, employer/nominated subcontractor agreement, form of nomination and subcontract agreement and conditions. The provisions are exceedingly complex and have been designed to cover the requirements of new construction works of some magnitude.

Agreement for Minor Building Works

This form is also issued by the JCT for minor building works for which a lump sum has been agreed and where an architect/supervising officer has been appointed on behalf of the employer. The heading to the form states that it is not suitable for use where a bill of quantities has been prepared, or where the employer wishes to nominate subcontractors or where the works or services are of a complex nature. However, in spite of its title, the form can be used for quite large jobs where the extent of the work to be done is reasonably certain and can be clearly defined so as to provide a realistic basis for a lump-sum tender. From the maintenance manager's viewpoint this is the most useful fixed-price contract format.

JCT Intermediate Form of Contract

This form is issued for contracts in the range between those for which the JCT Standard Form and the Minor Works Agreement are used and is suitable where the works involve the normal basic trades without complex specialist services. It requires the employer to provide at tender stage a set of drawings together with another document. Where this is a bill of quantities or schedule of work the contractor is required to price it, but where a specification he or she may supply a contract sum analysis. The conditions are less detailed than those in the Standard Form and in particular do not contain the complex provisions relating to the appointment of nominated subcontractors. Instead provision is made for work which is to be priced by the contractor to be carried out by a person named in the supporting tender document.

Standard Form of Management Contract

For works where a fixed lump sum is not required but an estimate of cost made prior to commitment. The fees for the management contractor are included in this form.

Fixed Fee Form of Prime Cost Contract

This is based on the standard form of building contract and requires the prior preparation of drawings and/or a specification on which the contractor bases his fixed fee for overheads and profit. It envisages new construction on a more extensive scale than most maintenance works and assumes that there is sufficient knowledge of the scope of the work at the time of tendering to produce a realistic estimate on which the contractor can base his fixed fee. On completion the contractor is paid the prime cost of labour, materials and plant as defined in the contract plus the fixed fee for overheads and profit.

Model Conditions of Contract for use with the national schedule of rates

This is modelled on the form that was produced for local authority measured term contracts. It differs significantly from the JCT forms of contract, which envisage single lump sum contracts. In particular it provides for a contract area and a contract period to be stated in the tender and for determination without default by either party at the end of six months from the date of commencement and thereafter on giving six weeks' notice. Orders for individual jobs during the contract period are to be in writing and should state a reasonable date for completion. The supervising officer is made responsible for arranging access to premises and contractors are entitled to recover the value of unproductive time should they be unable to gain access at the requisite time. Work is to be valued in accordance with the national schedule of rates or analogous rates or, where this would be unreasonable, as day work in accordance with the definition of prime cost of building works of a jobbing or maintenance character published by the RICS. The contractor is required to prepare his accounts for each and every order within 28 days of completion of the work, which is to be certified by the supervising officer within 28 days of receipt and payment made within 14 days of certification. Any defects that appear within three months of completion are to be made good by the contractor. Other provisions such as insurance against personal injury and damage to property are similar to those in the JCT minor works agreement.

Drawings

Drawings may in some cases be necessary to show the scope and extent of maintenance works, in conjunction with the written specifications. The standard conventions for the presentation of drawings are given in BS 1192 (Building Drawing Practice).

Specification

The presentation of information in a specification may take different forms according to the use of the specification and whether or not it is to be a contract document. For larger new works a specification may be prepared by the architect as part of the initial brief to the quantity surveyor to assist in the preparation of bills of quantities. For smaller new works such as alterations and extensions it is unlikely that there will be a bill of quantities and the contractor will be required to base his tender on the specification and drawings. For some types of maintenance work drawings may be unnecessary and the specification will be the sole source of information concerning the work to be done. In all cases, however, the information should be clear and unambiguous. Also care should be taken to ensure that statements about such matters as the nature of the subsoil are correct, otherwise the contractor may be able to rescind the contract or claim damages under the Misrepresentation Act 1967, or, if the statement has become a term of the contract, sue for breach of contract. Assistance in drafting clauses may be obtained from the National Building Specification or from the publication *Specification*.

A specification is normally divided into two main parts: preliminaries, and materials and workmanship.

Preliminaries

This section gives provisions that govern the general conduct of the contract and the overall extent of the contractor's liabilities. The clauses will have to be drafted to meet the requirements of the particular job and will include such matters as:

- general description of work
- form of contract to be used
- health and safety provisions under the construction (design and management) regulations
- provision of plant, scaffolding, storage, etc.
- water, lighting and power
- protective measures where work in occupied buildings
- times of access to site, restrictions on method of carrying out work, and other items likely to affect progress

Materials and workmanship

This section describes the quality of the materials to be used and the method of construction and standard of workmanship. For new works it is convenient to follow the order in which the work sections are given in the *Standard Method of Measurement of Building Works*:

- demolition
- excavation and earthwork
- piling
- concrete work
- brickwork and blockwork
- underpinning
- rubble walling and masonry
- asphalt work

- roofing
- woodwork (carpentry and joinery)
- structural studwork
- metal work
- plumbing and mechanical engineering installations
- electrical installations
- floor, wall and ceiling finishings
- glazing
- painting and decorating
- drainage
- fencing

Of course, the above list will have to be modified according to the nature and size of the job. For small works many of the work sections will not be required and also it may be advantageous to group two or more work sections together, e.g. excavation and concrete work.

For works in alterations and repairs it will often be necessary to depart from the sequence of work sections and adopt a sequence which follows the order in which the work will be carried out on site. In other cases such as internal decoration it may be more convenient, both from the point of view of preparing the specification and pricing, to group the work according to rooms in the order in which the rooms would be visited when walking round the building. However, assuming that the work section order is to be followed, the clauses within each work section would be grouped as follows:

- Clauses of general applicability to the work section.
- Materials and their preparation, application and protection (in some cases it may be better to group all the materials in a separate section to avoid repeating the description in each work section where the material is used).
- Work to be done.

The materials are described in the following ways:

- Giving a full description stating desirable and undesirable features and any tests with which they should comply.
- Stating the relevant British Standard Specification – note, however, that the BSS is usually the minimum quality and that in some cases more than one quality is provided for.
- Specifying a proprietary brand or naming a particular manufacturer or source of supply.
- Giving a prime cost (PC) sum and an outline description of the material – useful for such items as sanitary fittings in that decisions as to colour, etc., can be left until nearer the time of fixing.

Adjectives such as 'best' or 'first class' should be avoided unless they are recognised terms used to describe the particular quality required. Also the term 'other equal and approved' should be treated with caution. The workmanship clauses usually follow the order of carrying out the work on-site and should state precisely what is to be done and how, giving details of any constraints on the method of working, e.g. concreting to stop when the temperature falls below a certain level. Use can be made of BS Codes of Practice where appropriate and standard specification clauses for work of a repetitive nature. The right to select subcontractors for specialist work may be reserved either by naming the subcontractor or by including a prime cost sum to cover the cost of carrying out the work. Where a part of the work is too ill defined to describe accurately at the time of preparing the specification a provisional sum may be included to cover the estimated cost and an adjustment made to the contract sum on completion.

The type of specification described is called an operational specification in that it describes the actual physical work to be carried out by the builder. An alternative is the performance specification that specifies the performance to be achieved and leaves it to the builder to use his skill and knowledge to devise a physical solution that will meet the performance requirements. Thus, in the case of a central heating system one could state that the system is required to maintain a minimum temperature of 21°C in each room when the outside temperature is −1°C and leave it to the contractor to design the system. This places a greater liability on the contractor in that he or she is liable not only for the quality of the materials used and the standard of workmanship but also for the fitness for purpose of the system, i.e., that it meets the performance requirements.

Bills of quantities

Bills of quantities are normally only used for the larger jobs and where used in conjunction with the ICT standard form of contract must be prepared in accordance with the Standard Method of Measurement of building works. A bill of quantities sets out in a systematic manner the quantities and full descriptions of all the items of labour, materials and plant required to erect and complete a building. The preliminary section of the bill gives a brief description of the works and of the type of contract to be used and also details of general matters that may affect the price of the job. The measured work which follows is grouped into appropriate sections and each section starts with a set of preamble clauses which describe the quality of the main materials to be used and the standard of workmanship. Alternatively instead of giving the preamble clauses at the beginning of each section they may be grouped together in a separate section.

Usually the work sections follow the order in which they are given in the standard method of measurement but other groupings sometimes adopted are by functional elements, i.e. parts of the building which always perform the same function irrespective of the type of construction, or by operations, i.e. site activities.

The bill of quantities has two main purposes – firstly, to provide a uniform basis for competitive tenders by providing each contractor with an accurate statement of the quality and quantity of the work in a form suitable for pricing, and, secondly, to provide a schedule of prices for the valuing of variations ordered during the course of the contract. In addition the priced bills facilitate the valuation of interim certificates and the settlement of the final account and provide useful cost information for planning future projects.

Schedules of rates

These may be *ad hoc* schedules prepared for a particular job or standard printed schedules for use in term contracts. The *ad hoc* schedules usually follow the same pattern as a bill of quantities but may be either without quantities or with approximate quantities to assist in the evaluation of tenders. A standard schedule is one which has been designed to cover a range of repetitive jobs and may be one which is peculiar to a particular organisation or one which is more broadly based and of general applicability.

The National Schedule of Rates is produced jointly by the Construction Confederation and the Society of Chief Quantity Surveyors in Local Government[2]. It is a comprehensive computer-based system which has been designed principally for housing work with a maintenance bias. Each item in the schedule is broken down into labour and material elements so that it may also form the basis for bonus payments and materials allocation. The rates are updated quarterly to reflect the effect of inflation and changes in the pattern of prices. Also, the BMI *Price Book*[7] provides a useful schedule of the more common maintenance items and the rates are similarly broken down into their labour and material components.

The term schedule of rates is also used to describe the list of rates which have been used by contractors in building up their tender for a lump sum contract and which they are required to submit for the purpose of valuing any variations that might be ordered.

References

(1) Building Maintenance Information (1998) Review of Maintenance Procurement Practice. *Special Report Serial 270*. BMI, Kingston.
(2) *The National Schedule of Rates* (Published annually) (SCQSLG/BEC). BEC Publications, Birmingham.
(3) Audit Commission (1993) *Realising the Benefits of Competition*. Department of Environment, HMSO, London.
(4) Research Institute for Consumer Affairs (1971) *The Consumers' View of Building Maintenance*. Department of the Environment. HMSO, London.
(5) Orchard, R.R. (1966) The execution of maintenance – the contractor. Paper given at

Conference on Building Maintenance. Edinburgh, 1966.

(6) McCanlis, E.W. (1967) *Tendering Procedures and Contractual Arrangements*. Research Report, Royal Institution of Chartered Surveyors, London.

(7) Building Maintenance Information *Quarterly Price Book*. BMI Ltd, Kingston.

Chapter 9
Service Delivery

9.1 Maintenance managers as service providers

A maintenance management organisation can be seen as a service provider to the users of a building or estate, whose business it is to keep the building safe and functional with the minimum of disruption and disturbance. In this sense, maintenance management is as much about delivering a satisfactory service to people and organisations, as it is about the 'hardware' of the actual fabric of buildings and services. Often, users will judge a maintenance service on how it has directly affected them rather than on the quality of the repair carried out. Therefore, the degree of perceived client satisfaction with the overall service provided (rather than just the cost-efficiency of the repair and servicing work done) is a key indicator of service quality. This aspect of liaison with the occupiers and users is discussed in greater detail below.

At the point of delivery, maintenance and servicing works are carried out by the operatives and trades persons rather than the managers. The operatives may be employed directly by the maintenance management organisation (in the case of direct labour organisations), or, more commonly, by contractors sourced under the various procurement routes discussed in Chapter 8. Their levels of skills and experience, their motivation and incentives, and therefore their effectiveness in carrying out maintenance works, have received less discussion than is merited in view of the critical importance of their competence or otherwise in relation to the tasks they have to accomplish. The best managed and organised maintenance operation can founder on poor or inappropriate work by the operatives, yet there have been only a few serious attempts to determine exactly what skills and training are necessary to effectively execute works of maintenance and servicing in occupied buildings or to establish a framework for qualification in this area. This is dealt with in the section on skills and competencies of maintenance operatives.

9.2 Liaison with occupiers and users

The overall goals of a maintenance management organisation, discussed in detail in Chapter 2, usually entail striking a balance between needs and available resources, which may or may not involve a divergence of goals between the owners and users of the building (for example, in rented residential property). Thus it will rarely be the case that a maintenance management organisation has the resources immediately to fulfil every requirement and desire of the users. Yet it could be argued that the user's comfort and wellbeing is one of the key goals of any maintenance management organisation, within the proviso of limited resources. It follows that the 'people' skills of informing, listening, and explaining are important attributes a maintenance manager must cultivate within the organisation in order to be able to meet these user needs as far as possible, and to promote a positive image of the maintenance manager as an essential service provider. In the current business climate, the concept of customer service is paramount, and it is therefore important that the maintenance manager has regard to how the maintenance service interfaces with the occupiers and users of a building.

It can be seen from the above that what maintenance managers and their controlling organisation may regard as good practice (for example cost efficiency by bundling or prioritising repairs) may be perceived in a negative way by occupiers disadvantaged by such practice by having to wait for repairs which may seem to them to be essential. In a multioccupied estate such as rented housing or a large business organisation, any deferrals or refusals to repair may be interpreted by a particular occupier as inefficiency or favouritism in the absence of any more cogent explanation from the maintenance manager about cost-control and equity of maintenance policy across an estate.

Indeed the problem of negative perception of the building maintenance service by building users goes much deeper than this. People are only usually aware of the maintenance service when something goes wrong in a building. The occupiers are then faced with some form of delay (ranging from hours to years) whilst the repair is undertaken; the necessity of having to alter their schedules or go out of their way to make arrangements for the arrival of the operatives; the disruption and mess of the repair execution; and in many cases, the bill for the cost of the repair. At the end of this process there is apparently little to show except the restored function, often just as it was before. In some cases such as routine servicing, the disruption caused has not even the perceived benefit to the occupier of restoring a malfunction, since the purpose of the operation is to prevent any malfunction in the first place. This apparent paucity of positive psychological reward in return for what may have been a lengthy, costly and aggravating episode may lead users to externalise this frustration on the perceived performance of the maintenance organisation itself. Hence some maintenance managers may come to feel that this negative image is intrinsic in the job, and that there will be little if any praise and some level of criticism whether the repair operations are done well or not. Such fatalism may itself have a very negative effect on the quality of maintenance service provision by insulating the manager against any criticism, whether justified or not. It is better for the maintenance manager to recognise that occupiers may tend towards a negative perception of building maintenance for the reasons stated above, but that this can be lessened or reversed by attention to the following quality indicators:

- actively seeking the occupier's opinions and views
- keeping the occupier informed on the 'what, when, why' – what is happening, when it will happen, and why it is happening
- focusing on the quality of the repair operation on site:
  o quality of appointments system: flexibility, and whether appointments are kept by contractors
  o skills and qualities of the operatives: technical skills, and 'people' skills
  o health and safety safeguards for the occupants
- regular monitoring of the building environment to anticipate shortcomings and malfunctions.

9.3 Attributes of maintenance service quality

Looked at from the occupier's point of view, the important attributes of maintenance service quality can be identified as:

- Reporting of repairs
  o Are wants of repair routinely monitored, for example by condition surveys, or do the occupiers have to report all breakdowns and wants themselves?
  o How easy is it to report a want of repair?
  o How are occupiers kept informed of the progress of their repair requests, and how can they influence the progress to suit their own needs and agendas?
  o If a repair is only a low priority and not scheduled for immediate action, how is this explained to the occupier?
- Carrying out repairs
  o Are appointments to repair carried out at a time to suit the occupier?
  o Are the repairs carried out quickly and efficiently?
  o Are the repair operatives polite and helpful?
  o Are the repairs carried out with due regard to the health and safety of the occupiers?
  o Are the areas affected adequately screened, and any mess properly cleared up on completion?
  o Is the occupier's opinion on the quality of the service sought and valued?
- Maintenance standards generally
  o Are the occupiers consulted about the quality of their environment?

    ○ Are any objective measures of building performance and building environment collected by the maintenance management organisation?

    ○ Are the occupiers empowered to carry out any works or alterations themselves if they wish, and under what circumstances?

    ○ What input to the decision-making process do occupiers who are directly paying for the service make? (For example, occupiers in leased premises paying a service charge for services over which they have no direct management or control, but which affects their occupation and enjoyment).

Reporting of repairs

Repair notification is usually initiated by building users and occupiers, and the responses of the maintenance management organisation will be a critical determinant of perceived service levels. The first issue is how easy it is to report the repair need. A frequent criticism of large bureaucratic maintenance management organisations has been the impersonal nature of this 'reception' function, and indeed the obstacles that may be perceived to lie in the path of the reporter to actually log the repair request in the first place. For example, referring telephone callers between various extensions, diverting them into call-waiting and 'voice-jails' if the service is understaffed and oversubscribed are frequent sources of frustration which can sour the user's perceptions of service quality at the very outset. So the first requirement is to have a single and well-advertised point of access to the repairs and maintenance service, often known as a helpdesk, where the caller will be answered promptly and courteously, and where the information required from them to record the request will be taken, and explanations of the process given. This helpdesk may be located centrally, to deal with telephone callers, or may be more local to the users, for example in estate offices. The helpdesk operators need not be the maintenance technicians and managers themselves but preferably administrative staff better-suited to this 'people' skill who can interact effectively with the caller, whatever the caller's background. The administrative staff will obviously need some form of basic technical training to be able to process the repair request in sufficient detail to enable them, or the technical staff, to prioritise and make decisions about action. This technical training will increasingly rely on the use of dedicated information software which guides the repair recording procedure. The key attributes of repair logging are:

- name of caller, telephone number, and details of when available
- address/location of the repair
- nature of the repair requested
- preferred access times
- specific relevant information, such as presence of young children or elderly persons who may be adversely affected by the breakdown.

According to the nature of the repair reported, which may range from the critical (a health and safety hazard such as loose exterior cladding or a breakdown in sanitation) to the trivial (a squeaking door), the caller should be informed (at the time of their first call if possible) about the priority of the repair, and the timescale (hours, days, weeks, if and when) in which it is likely to be undertaken. In the case of low-priority repairs, the helpdesk operative may need to explain why it has been assigned a certain priority.

It is important to have a follow up or review procedure for repair requests received. Occupiers who report a repair and are then left with no indication of what, if anything, is being done, or who are promised action or information by a certain time which then does not happen, are understandably likely to form a poor image of the service. In some cases this can be serious enough to generate its own considerable workload and costs in dealing with callers re-reporting repairs they feel may have become lost or neglected within the maintenance management system. More seriously, some local councils who have not had the resources to action all necessary repair requests, have found themselves in a 'litigation culture' whereby a tenant seeks legal action by housing notice or environmental health notice (cf. Chapter 3) as a first resort rather than a last, with the subsequent drain of even these limited maintenance resources into the legal process rather than the repair process.

Carrying out repairs

When undertaking repairs it is important that the occupiers and users are consulted as to when the work should be done, and in some cases to have input into how the work is done in relation to interrupted services, access, or space use. In terms of when the work is carried out then an appointment will usually need to be made, usually by the contractor or DLO directly, though the efficiency of the appointments service must be monitored by the client maintenance management organisation. Appointments once made should be kept wherever possible as a matter of policy, and this should be carefully monitored as a key service quality index by the maintenance manager. Appointments missed without apology or explanation are a major cause of dissatisfaction, so if for unavoidable operational reasons an appointment cannot be met, it is incumbent to inform the affected occupiers as quickly as possible so they are not waiting for a visit which will not happen. The subsequent reappointment should be made a priority to prevent any further inconvenience to the occupier. In commercial premises normally occupied during business hours, it may be sufficient to specify just the day or even the week when works are to commence depending on how much the work will interfere with the normal operation of the building.

For residential properties, the occupier may be put to considerable inconvenience in attending a repair, by for example having to take time off work or make special arrangements for taking children to school, etc., and therefore the consequences of a missed appointment are going to be much more seriously felt by them than by the contractor. To improve levels of service in this respect, some maintenance management organisations have extended the times repairs are carried out until 8 PM to cater for the difficulty some working families may have in meeting daytime appointments[1].

When the maintenance operatives are on site, it is important that they conduct themselves in a manner that respects the occupiers and their environment. Language and behaviour that may be acceptable on a building site may cause deep offence to certain occupiers, particularly in relation to obscenities, sexual, political, or racial remarks. In addition, the operatives must realise that by working in someone else's environment, it is incumbent on them to behave as a guest and where possible, to respect the occupiers' standards of housekeeping and cleanliness in particular. This is especially important in relation to residential premises, for which occupiers are likely to have much deeper feelings than for their workplaces.

The inevitable disruption that maintenance works can generate is a source of stress for occupiers, particularly in residential buildings. The specification and practice of effective screening, and rigorous cleanup routines at the end of each working day, and thoroughly on completion, is therefore an essential component of good maintenance work. Where some mess is unavoidable, for example during lifting floorboards and cutting chases when rewiring, occupants should be informed in advance so that delicate or valuable goods or items of furniture can be moved or protected. In the case of elderly occupants or those with disabilities some direct assistance will often need to be given in this respect. The interruption of essential services such as water, sanitation, and electricity should similarly be discussed with the occupant in order to come to a satisfactory arrangement, and in any case such 'down time' should be kept to the minimum, even where this may impose extra costs on carrying out the work. Some cases may necessitate the maintenance contractor/DLO providing temporary services where, for example, there may be installations requiring uninterrupted service, such as heating in elderly peoples' homes. For major repairs such as dry rot treatment, it may be necessary to evacuate the entire property for a period of days, in which case arrangements for moving the tenants to temporary accommodation for the duration of the work and organising security may take as much time and effort by the maintenance manager as the actual work itself. Finally, in rented residential accommodation, the repair works may unavoidably cause some degree of damage to decorations, which although the responsibility of the tenant, will require either reinstatement work by the contractors themselves; or, more usually, decoration vouchers or a rent holiday being given to the occupier by way of compensation.

Health and safety whilst working in occupied premises is very important. The maintenance manager and operative must have regard not only for their own health and safety, but also that of the occupier. Despite the health and safety legislation, notably the Control of Substances Hazardous to Health (COSHH) Regulations (cf. Chapter 3), it is still common to disregard the occupants in any assessment. For example, most contractors will be accustomed to routinely preparing health and safety and COSHH assessments under the Construction (Design and Management) Regulations, but close scrutiny may reveal that the assessments only cover the health and safety of the operatives themselves. Whereas good working practice will in any case minimise the exposure of bystanders, there may be cases where, for example, operatives are wearing required breathing apparatus to protect against dust inhalation when nearby occupiers are not. Obviously in such cases, effective screening takes on a greater importance than just to minimise mess and disruption – it may be essential to separate the occupiers from potentially harmful processes and operations. In addition, particular occupiers may be potentially more vulnerable to health risks associated with building work. For example, children with a predisposition to asthma, common enough these days, may be adversely affected by dust generated at far lower thresholds than the adult maintenance operatives for whom the COSHH assessments have been prepared. It is therefore necessary to have a formal management routine to identify possible health risks to the occupiers at the start of maintenance works and take appropriate steps to minimise the risk, and this responsibility (usually on the contractor) should be specifically featured in the contractor's health and safety statement.

Occupier feedback

As was stated at the beginning of the chapter, the quality of the maintenance service relates not only to the standard of repair, but to the occupier's perception of the service provided. In this respect it is desirable to monitor the occupier's experience. This can be done at the lowest level by having a complaints procedure in place, which is a formal channel by which perceived poor service can be reported and judged. This procedure should entail a formal acknowledgement of any complaint received, and a commitment to inform the complainant what if any action has been taken as a consequence. At a higher level, a mechanism to routinely gather the occupiers' opinions about the repair and maintenance service may be instituted as part of a wider quality control system, for example as would be required under ISO 9000 Quality Assessment. One of the simpler and easily managed methods is the use of a reply-paid postcard, left by the operative on completion of a repair, which asks the occupier to rate the service under a number of headings. The headings may include:

- apparent quality of the repair (is it satisfactory as far as the occupier is aware?)
- convenience of appointment
- whether the appointment was kept
- politeness/helpfulness of the operative
- whether the clean-up was satisfactory.

Space can be left to pick up any other comments the occupier may have. This postcard system is becoming widely used by social landlords, and has several benefits:

- it provides a convenient vehicle for complaints where warranted
- it provides routine feedback on contractor performance
- it enables problem areas/contractors to be targeted for closer supervision
- it empowers the occupiers, who feel that their opinion is sought and valued
- it enables positive feedback (i.e. good service) to be noted as well as negative.

Maintenance standards generally

For occupiers in commercial premises, or for feedback on the maintenance service generally (rather than specific repair instances), questionnaire responses may be sought. The questionnaires may also focus on perceived standards of accommodation and suitability for the user's needs. In the broader context of facilities management, this is known as a post-occupancy evaluation (POE).

Post-occupancy evaluation

There are two basic methods of evaluation, known as ORBIT and Building Quality Analysis (BQA). The latter, developed by Bruhns and Isaacs[2] involves the assessment and grading of a building in nine categories:

(1) *Presentation.* The aesthetics and impact of the building on occupiers and users.
(2) *Space.* The effectiveness of space utilisation. Note that too much space is wasteful not only in terms of building fabric, but also in the unnecessary provision of heating, lighting, and other services to wasted space. Too little space interferes with the use of the building.
(3) *Access and circulation.* This concerns an examination of the effectiveness of flows of people, goods, and information around the building. It evaluates signing and route finding, and journey times around the building.
(4) *Business services.* This is concerned with the evaluation of services such as electricity, computers, mail, and telecommunications provision.
(5) *Personnel amenities.* This looks at washing and sanitary facilities, restrooms, cafeterias, staff parking, etc.
(6) *Working environment.* This examines the ergonomics of the workspaces and their suitability to individual uses and needs.
(7) *Health and safety.* This looks at possible health and safety hazards and the protection against them.
(8) *Structural considerations.* This assesses the suitability of the structure for the use, and its durability and maintenance characteristics.

(9) *Manageability.* This specifically assesses the building for ease of cleaning, of maintenance, and general running.

Obviously not all of the above functions fall within the remit of the maintenance manager, however, the characteristics of the building in terms of maintenance and servicing are important components in the assessment of the quality of the building. The effectiveness of the maintenance regime will have a significant impact in several areas.

The BQA is carried out in a three-stage process, illustrated in Fig. 9.1. Firstly, a task group representing the managers and users of the building, together with the quality assessors themselves, is formed. This group should be less than 12 persons to ensure manageability in terms of meetings and decisions made.

The initial stage is the *Walkthrough.* The task group tours the building and talks to the individual users, in order to get a focus on what are the significant

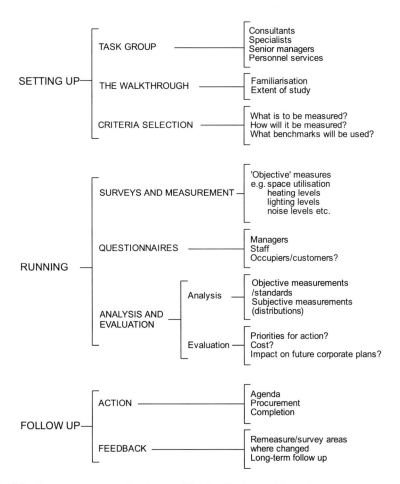

Fig. 9.1   Post-occupancy evaluation model (after Bruhns and Isaacs).

issues and parameters they will go on to study in more detail. For example, if a number of users volunteer a complaint of sore throats and headaches during the walkthrough, it would direct the task group's attention towards a more detailed evaluation of the air conditioning system as a possible source of sick building syndrome. After the walkthrough a task group meeting decides the detailed criteria for the in-depth study to follow.

During the second stage, the *Running* of the main evaluations, information is collected in two ways. Firstly, objective measurements are taken of relevant parameters of the building, for example, temperature means and variations, amount of floor space per user, or lux levels of task lighting. Secondly, the users are questioned about their own perceptions of the building and their particular environment. This questionnaire must be structured very carefully to avoid bias towards the negative, always a trend when people are asked their opinion. The structure seeks to achieve a balance to determine what aspects of the building are satisfactory, what may be overprovided, and what are perceived as below par. The most effective vehicle to obtain a useable response from the occupiers which will both reflect their personal viewpoint (for example, at a given temperature, some will find it too hot; others too cold), and permit collation and interpretation of the data, is the use of bipolar scales as the medium for answering questions. This entails the user indicating an opinion on service provisions on a scale from 1 to 10, with 5 being neutral, as indicated in Fig. 9.2.

The advantage of this system is that the user can give a proportional and quantitative response to a question (instead of a straight yes or no, or written opinions, which are difficult to process in bulk to arrive at meaningful information); and these responses may be taken together per floor, section, department, or for the building as a whole, and data processed to give a distribution of responses, as indicated in Fig. 9.3.

The position of the mean point on the scale will indicate on average whether the provision of the service in question is perceived as good, bad, or indifferent. A wide standard deviation would indicate a wide spread of opinion (for example on lighting levels or temperature), and would indicate on analysis that one standard throughout the building would be unlikely to be satisfactory. In such a case, introducing some degree of local user control would be indicated as a worthwhile remedial measure. Grouped together, the responses to the various questions can be graphically illustrated as shown in Fig. 9.4.

This indicates across a range of provisions the mean level of satisfaction, and the variance around this mean. Coupled with the assessment of perceived

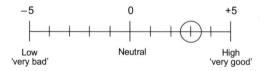

Fig. 9.2   Bipolar scale for occupier questionnaire.

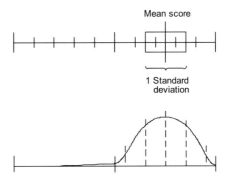

Fig. 9.3   Spread of responses: standard deviation from the mean.

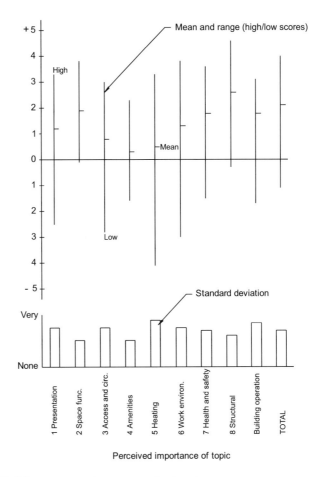

Fig. 9.4   Building assessment – ranges and categories.

importance of the question to the user, indicated at the foot of Fig. 9.4, this may be input into the final stage of the POE: the *Follow Up* evaluation. In this stage, the objective and subjective results are analysed to determine:

- the perceptions of importance for each of the subject areas
- the degree of satisfaction for each
- the range of satisfaction for each.

Note that high levels of satisfaction for items perceived as relatively unimportant may indicate overprovision in these areas; and resources can be diverted from these areas to other more important and/or underperforming areas. A fine-tuning strategy may then be formulated for the building depending on the seriousness of the findings and available resources.

Whilst the full remit of a POE does include many areas not of direct relevance to the maintenance manager, it can be seen that the quality of the maintenance service provision will have a significant impact on many of them. In particular, those areas related to the building fabric (ease of opening windows, waits for lifts, state of floor coverings, decorations, etc.), and to building services provision (heating, ventilation, lighting, electricity, communications) will reflect in part the efficiency of the maintenance regime.

User participation in management

Other issues relating to the general standards of maintenance service provision include the degree of occupiers' empowerment relating to what they may do for themselves to their built environment. Too restrictive a regime, particularly in relation to residential premises, may discourage the occupiers' identification and involvement with the premises, and lead to a negative viewpoint. It is usual therefore to permit tenants' alterations provided suitable safeguards are met (such as planning and building control permission being obtained, approved contractors used for certain classes of work, etc.), and that the work may require reinstatement on termination of the tenancy. In this respect the system's operation may be essentially the same as the legal position regarding alterations, dilapidation and the tort of waste in leasehold law.

In some public sector and voluntary maintenance management organisations, the entire organisation itself may be subject to some level of control by tenants and occupiers. For example, some local authorities such as Wirral Metropolitan District Council have devolved the entire day-to-day maintenance management for some of their estates on to self-managed tenant co-operatives, with a degree of success. Liaison and advice is provided on request to the co-operative by dedicated council officers, but repair priorities and regimes are decided and undertaken by the tenants' groups themselves. Nearly all social property owners now have tenant representatives on their boards of management. Many business organisations have management committees drawn from various business areas, which input at board level into the maintenance management function. It is clear that across the property sector, the trend away from top-down, corporate building maintenance management structures and towards the involvement and empowerment of tenants and occupiers in building maintenance management is continuing.

In sum, the maintenance management operation should seek to be receptive, unobtrusive, and effective from the occupier's point of view. Barrett[3] highlights the key index of service quality in the related field of facilities management as the difference between the service perceived by the occupier and the service expected. He states, 'the facilities manager should be careful not to overstate what the facilities department is capable of delivering. If this occurs, it is obvious that the client is unlikely to be satisfied with the service provided. However, in a similar vein, it is important to portray a positive, rather than negative image.' He goes on to point out that a good level of service may merely pass unnoticed: 'Quite simply, they (the facilities management organisation) have avoided problems and, as a result, their work may appear to be deceptively simple to the rest of the organisation.'

9.4 Skills and competencies of maintenance operatives

In this section, the nature of maintenance work and the consequent skills required of maintenance operatives for the delivery of an effective service will be examined.

Maintenance is a labour-intensive activity and therefore the greatest economies are likely to flow from measures that improve labour productivity. Studies[4] have shown that in the case of painting the labour content may be as high as 85% of prime cost and for general repair work approximately 65% of prime cost. The small scale of much of the work and the restricted conditions under which it is executed preclude the extensive use of mechanical aids. About 40% of all construction operatives are employed on maintenance and repair, in the main by small local firms. This does not include operatives directly employed in commerce and industry, and taking these into account it is probable that the total manpower employed on maintenance and repair is almost as numerous as that employed on new construction.

The trades employed in repairs and maintenance are effectively the same as the broadly based construction trades and many small contractors use the same operatives for both types of work. The five main trades (carpenter, bricklayer, plasterer, plumber, painter) account for 80% of the maintenance and repair labour force, the predominant trade being painters, who comprise some 30% of the total. Indeed about 75% of all painters and decorators are employed on maintenance and it would be beneficial if their training were directed towards this type of work. Table 9.1 gives a general picture of the composition of the labour

Table 9.1 Labour force composition.

Trade	Small contractor (less than 50 staff) %	National contractor (maintenance dept.) %	Local authority (direct labour dept.) %
Carpenter	25	26	16
Bricklayer	22	11	10
Plumber	4	9	10
Plasterer	1	2	6
Painter	26	39	34
Others	2	1	4
Handyman	1	1	2
Labourer	19	11	18
	100	100	100

forces of the three main types of organisation concerned with the execution of the maintenance work.

The higher proportion of bricklayers employed by small contracting firms is a result of the fact that such firms undertake a certain amount of minor new works in the form of alterations and extensions and a lower proportion of painting and decorating than the other two types of organisation. It is usual for maintenance operatives to work in small gangs made up of a number of different trades. The BRE study of operatives' work in this field[5] revealed a good deal of overlapping between trades and also suggested that the relationship between a tradesperson and the corresponding trade labourer was less clearly defined than in new work. The study revealed a 'large area of general maintenance and common work which defied normal trade classification and which was performed indiscriminately by most trades and their associated labourers'. In the years since this study was carried out, it would appear little has changed. A report from the Joseph Rowntree Foundation on housing maintenance[6] showed that in small builders carrying out maintenance works, 50% of operatives held a City and Guilds craft certificate, a further 6% had served an apprenticeship, but 44% had no formal qualification at all.

However, it is clear that rigid trade demarcations can hinder the economic carrying out of maintenance work. To quote from the evidence submitted to the Phelps Brown Committee by the Institute of Municipal Building Management (now the Institute of Maintenance and Building Management), 'The encouragement of multicraft skills could play an important part in controlling the rising cost of maintenance, by the introduction of a "building trade technician" recognised by the trade unions as an operative who in day-to-day maintenance is permitted to undertake work in more than one trade'.

It is generally recognised that maintenance requires greater skill than new construction work and the previously mentioned study of operatives' work demonstrated that in fact maintenance operatives were better qualified than those in new construction. In maintenance about 50% of operatives under 30 years of age had a craft certificate and six out of ten of these had achieved advanced craft certificate level. The corresponding proportions in new construction were 38% with a craft certificate, three out of ten having gained an advanced certificate.

Unfortunately, the wage structure laid down by the National Joint Council for the Building Industry is based on a simple division between craftspeople and labourers with only a limited number of plus rates in between for certain semiskilled operatives such as steelbenders, scaffolders, plant operators, etc. In maintenance at least, it could be argued that there is a need to formulate a pay structure which will reflect the gradations of skill involved, so that the versatile craftsperson can be more adequately rewarded. Generally plus rates, i.e., payments above the basic wage rate, are very much higher for new construction than for maintenance and are unrelated to skill and, in many cases, to productivity. There is little recognition of the fact that in maintenance there is an emphasis on an operative's responsibility and initiative which often requires the operative to work without much supervision and with a minimum of instruction and technical guidance.

Age and stability

The labour force in maintenance and repairs tends to be older and more stable than that in new construction works. The aforementioned study of operatives' work revealed that the proportion of those over 40 years of age was 58% for maintenance as compared with 32% for new construction. Since the proportion of apprentices is higher in maintenance than in new construction, there would seem to be a drift from maintenance to new construction shortly after the completion of training and a reverse movement after the age of 30. This is most pronounced for bricklayers and plumbers and to a lesser extent for carpenters and painters. The reason for the reverse movement of craftspeople is probably quite simply that as they become older and acquire family responsibilities, the greater stability, local nature and less rigorous working conditions in maintenance outweigh the higher bonus earnings obtainable in new construction. This seems to be borne out by the finding that only 25% of operatives employed by small firms associated with maintenance were attracted to other jobs by incentive payments, as compared to 59% for large firms mainly concerned with new construction. The emphasis placed on stability of employment is brought out even more clearly in the case of local authority direct labour forces which usually have a much lower labour turnover even than small building firms. Other factors which attract the older operative to maintenance may be the greater responsibility and sense of satisfaction in the work done than is usually possible in new construction, coupled with the greater variety of maintenance jobs and the personal contacts with building owners and tenants.

While the above would suggest that maintenance operatives are not so highly motivated by financial reward, it is more probable that the deciding factors are the disincentives of working on new construction sites. If these disincentives were removed by changes in employment policies to give greater continuity of employment and better working conditions, the drift to maintenance in later life might well be restricted. The balance could be preserved by making the rewards and conditions in the maintenance sector more attractive so as to produce a corresponding reduction in the movement of the younger craftspeople to new construction.

Multiskilling: the maintenance technician

It is generally conceded that it would be desirable to have a specially trained maintenance operative. A practical difficulty in introducing a separate category of 'maintenance technician' is that it would cut across the traditional methods of grouping craft skills. The Association of Direct Labour Organisations (ADLO) has been one of the few bodies to publish guidance on what form this might take[7].

The present groupings are:

(1) *By material.* Traditionally trades are associated with a particular material, e.g. plumbers with lead, carpenters with wood, etc.
(2) *By tools or plant.* This is really complementary to the first grouping in that special tools have been developed to shape and fix particular materials. However, dexterity in the use of a particular tool may enable a craftsperson to undertake work outside his or her normal field, e.g. a trowel is used for both bricklaying and plastering and, therefore, a grouping of 'trowel trades' would cover both of these activities.
(3) *By function.* There is a trend for trades to identify with the provision and maintenance of a functional element or system rather than the traditional material, e.g. plumbers with water services, although it is now usual for materials other than lead to be used.
(4) *By operational method.* This grouping has emerged in response to the need for special skills for the erection of industrialised building systems. In a number of cases these methods are unique to the particular system and are not based on the traditional craft skills.

The above groupings are far from ideal from a maintenance point of view. For this purpose, it would be more useful to have a grouping based on the work to be carried out at a particular location, e.g. piecing in a new section to a defective window frame, including carpentry, painting, making good plaster reveals and perhaps reglazing.

This would raise the question as to whether a maintenance technician should be treated initially as such or should have additional training after acquiring skill in one of the main trades. This can be expressed as the relationship between depth and breadth in multiskilling qualifications (Figs 9.5 and 9.6).

The 'depth' of skill relates to the degree of competence in a particular skill or trade. The 'breadth' relates to the spread of competencies across different trades (for example, a bricklayer who may, at a lesser level of skill, be capable of simple tiling), and to more general competencies of reading drawings, liasing with building occupiers, etc. Clearly too much breadth without sufficient depth of skill would lead to a 'jack of all trades' capable only of mediocre skills, albeit across a range of trades. On the other hand, too much depth may mean the operative has redundant levels of skills little used in maintenance works and therefore would be of limited use in carrying out such works. For example, a sink replacement requires six trades – plumbing, carpentry, tiling, plastering, painting, and electrical – and it would obviously be grossly inefficient to send six tradespeople for this one job. Obviously, a plumber with some training in basic carpentry, tiling, and plastering techniques would be of far more use in building maintenance works. However, having said this, there are some trades, notably gas fitting and live electrical work, which do require a dedicated 'depth' tradesperson because of the importance of health and safety requirements in these areas.

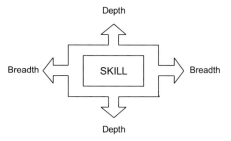

Fig. 9.5   Skills template.

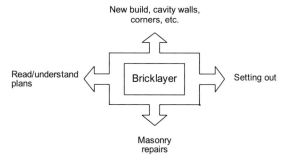

Fig. 9.6   Bricklayer skills template.

One effect of creating a specialist maintenance technician would be to restrict the free movement of operatives between new construction and maintenance. This might be thought disadvantageous in that the movement from one construction activity to the other provides a ready means of adjusting to fluctuations in the demand for new construction. However, this situation accounts for the unfortunate 'casual' nature of employment on new construction sites and also results in neglect of the existing stock of buildings and higher maintenance costs when there is a heavy demand for new buildings. It is in fact a consequence of an imperfect system and could be avoided by rationalising the demand for new construction so that continuity of work is provided for a relatively stable labour force. Indeed, the greater industrialisation of building involving high capital investment in equipment will depend to a large extent for its success on an assured market. In addition, industrialisation has led to a divergence between the skills required for the initial fabrication and assembly and the subsequent maintenance. This is seen most clearly in the case of manufactured products such as cars and is likely to apply equally to factory-produced buildings.

The ensuing polarisation of the construction industry will emphasise the need for specially trained maintenance technicians and a separate wage structure commensurate with their skills and responsibilities. Additionally, the formation of single-purpose organisations concerned solely with maintenance would foster the development of more sophisticated planning techniques for the grouping of work so as to achieve economies of scale. In recent years there have been some encouraging signs that this is beginning to happen. The most significant is the development and delivery of a new National Vocational Qualification (NVQ) in Maintenance approved by the Construction Industry Training Board (CITB)[8]. The qualification uses trades modules from existing NVQ courses, from which the maintenance technician takes any 3 from 12. Thus the 'multi-skilling' still recognises necessary trade divisions and competencies whilst acknowledging that for the maintenance operatives, a combination of three related lower-level competencies actually needed to execute maintenance works is of far greater relevance than an in-depth knowledge of only one. As well as the specific trade skills, the course requires core competencies in six general areas of particular relevance to maintenance works. In outline the proposed structure is as follows:

Core skills (mandatory)
(1)   contribute to the provision of efficient working practices
(2)   handle, store, and secure resources
(3)   develop and maintain a positive relationship with customers
(4)   record and maintain information (site measurements and surveys)
(5)   erect and dismantle access platforms (basic working platforms)
(6)   complete works for handing over

Option units (any 3 from 12)
(1) plumbing (basic)
(2) carpentry
(3) plastering
(4) wall/floor tiling
(5) painting and decorating
(6) bricklaying
(7) roof slating and tiling
(8) glazing
(9) concretor/pavior
(10) built-up felt roofing
(11) gas installation
(12) electrical (basic)

To date the qualification is in its infancy, but it is hoped that with the Department of Environment and Transport's current initiatives (notably the Construction Skills Certification Scheme, or CSCS) to foster a registration scheme to protect the public against so-called 'cowboy' builders, its increasing relevance to maintenance work will be recognised and supported.

Operatives' productivity and output

Having considered the skills required of the maintenance operatives, the factors which affect the levels of working efficiency and reasons for the relatively low productivity output in maintenance will be discussed. These are:

(1) The small scale of the individual jobs, resulting in a high proportion of unproductive time. The time spent in collecting stores, preparing to do the work and clearing up on completion is substantially the same irrespective of the size of the job.
(2) The diversity of the work content of the jobs. Even quite small jobs can involve a number of separate trades, leading to delays and additional cost if demarcation rules are strictly adhered to.
(3) The nonrepetitive nature of much of the work. Each job presents a fresh set of problems and may require novel and unfamiliar methods of working.
(4) The adoption of inappropriate methods. This follows from the previous point in that the lack of continuity does not permit the development of optimum methods. The effect in this case is not limited to increasing the cost of the particular operation, but may also have an adverse effect on the long-term durability of the building and the cost of subsequent maintenance.
(5) The dispersal of sites. The proportion of unproductive travelling time is related to the size of the job and the distance between site and depot. Also, the higher the dispersion factor the more difficult and expensive it is to provide effective supervision. An additional factor, particularly in relation to

housing maintenance, is the time wasted on abortive visits through failure to gain access to the premises.

(6) Difficult working conditions arising from:
- need to work in confined positions
- inadequate means of access
- restrictions on working times or methods
- protective and safety measures necessary in an occupied building
- inclement weather in the case of external works.

(7) The making good of damage caused by the removal of old components and installations and the extra care necessary to limit consequential damage to a minimum.

(8) Poor communications between the client and the contractor, failing to keep the client fully informed of progress.

(9) Limitations on the use of mechanical plant imposed by the small scale of the operations.

(10) The tendency to execute work only when it becomes a matter of urgency. This not only has a disruptive effect on the organisation of the labour force, but may result in deterioration of adjoining elements necessitating much higher expenditure than would have been necessary if the defect had been remedied earlier.

(11) The difficulty of drawing up a detailed programme of work before a fairly extensive exploratory survey has been carried out to determine the extent of the damage, e.g. dry rot.

(12) Obsolete or obsolescent materials that may be required to match existing work and which become increasingly expensive.

(13) The poor management quality of many small firms engaged in this type of work and the lower calibre of professional support.

(14) The age distribution of the operatives. It has been noted that older crafts-people tend to gravitate to maintenance because they are attracted by the slower tempo of the work, while at the same time firms engaged in this work employ a higher proportion of apprentices.

(15) Motivation. Many different theories of motivation have been proposed but none is entirely satisfactory. Hertzberg proposed that workers had two sets of needs, those which when fulfilled promoted job satisfaction and those which when unfulfilled produced dissatisfaction. The former, which he described as motivators, included achievement, recognition, the work itself, responsibility and advancement. The latter, which he described as hygiene factors, were largely to do with the working environment and included company policy, relations with supervisors and working conditions. He found that the factors that caused job satisfaction were quite separate and distinct from those that caused dissatisfaction. Thus, an improvement in hygiene factors, whilst removing a source of dissatisfaction, does not necessarily act as a motivator. Maslow postulated a hierarchy of needs as illustrated in Fig. 9.7 and that needs only act as motivators when they are

Fig. 9.7 Maslow's hierarchy of needs.

unsatisfied. Thus, the lower physiological needs are dominant until satisfied and then the next lower becomes dominant, and so on.

Physiological needs

Greater job satisfaction can be provided by job enlargement. This may take the form of horizontal job enlargement in which the worker is involved in a wider range of operations at the same level in the hierarchy or vertical job enlargement, also called job enrichment, in which certain aspects of the supervisory function are delegated to the worker. Participation may also be regarded as a form of job enlargement in that workers are involved in discussions on such matters as standards and methods of working. It is important in such situations that the individual is given sufficient information to allow him to participate effectively.

Incentive schemes

An obvious way to improve output would appear to be the provision of some sort of incentive to work harder. The objects of incentive schemes are stated in the National Working Rules for the Building Industry as:

(1) To increase efficiency, thereby keeping costs at an economic level.
(2) To encourage greater productivity, thereby providing an opportunity for increased earnings by increased effort while maintaining a high standard of workmanship and avoiding a waste of labour and materials.

However, one should examine the reasons for attempting to increase efficiency and particularly, as far as direct labour is concerned, whether the intention is:

(1) to do more work with the same labour force, or
(2) to do the same work with a reduced labour force, or
(3) to deal more quickly with work orders.

It is also clear that in some cases increased wages are paid for the purpose of attracting and retaining operatives and any increase in output that results is quite fortuitous. Local authorities are at a particular disadvantage when competing with contractors for scarce labour in that, unlike contractors, they are unable to pay 'plus' rates. Any additional payments must, therefore, be made under the guise of an incentive scheme. From the contractor's point of view, the value of incentive schemes will be judged by the effect on profits. As many small maintenance jobs are let on a 'day work' or 'cost-plus' basis (i.e. the contractor is paid the actual cost of labour and material plus a percentage to cover overheads and profit), there is apparently no gain from carrying out the work more economically. Indeed, it could be said that the more inefficiently the work is executed, the higher the cost and the greater the contractor's profit. In a survey carried out by the Advisory Service for the Building Industry[9] some firms admitted that they were not interested in improving productivity on day work where their profit margins are assured. It was remarked that although it is against the interest of the general public, this mistaken attitude is probably widespread.

Incentive schemes may be broadly divided into those that provide a direct incentive by setting targets for the execution of separate parts of the work and those that provide an indirect incentive by attempting to involve operatives in company policy.

Direct incentive schemes

(1) *Piecework.* The target is a lump sum payable for completion of a specified task irrespective of the time taken. It is thus possible for an operative to earn less than the standard basic wage and the scheme is usually confined to labour-only subcontracting in new construction work. The view has been expressed that cash targets do not have the same incentive effect as hourly targets and that they are probably more suitable for the smaller than the larger firms. While it is difficult to reconcile this statement with the rapid growth in recent years of labour-only subcontracting, it may be true in respect of the older operatives forming the bulk of the maintenance labour force.

(2) *Spot bonuses.* These are offers of additional sums of money over and above the basic rate and are awarded mainly by working principals for urgent work. By their very nature they are not applied continuously and therefore have little effect on long-term productivity.

(3) *Targets related to estimate.* Here the targets for the various operations are derived from the rates used in building up the estimate. The major defect of this system is that where the estimates are competitive the rates will fluctuate according to market conditions and will not provide a uniform basis for output-related earnings. However, if the rates are adjusted the scheme can be comparatively cheap and simple to administer and can provide a continuous check on profitability and a good basis for labour control.

Usually a percentage is deducted from the estimated time to create a reserve during favourable times to counterbalance losses during unfavourable times. A problem associated with this method is that in practice the labour content of the rate is based on experience, perhaps aided by past records of the time taken to perform similar work, and in a number of cases it is little more than a guess. This is particularly so where the estimator is remote from the site and is unable to check the methods of working. In such cases the estimator proceeds by trial and error – if the workers complain that they are unable to earn bonuses, he or she assumes the target is too 'tight', while if excessive bonuses are earned the assumption is made that the targets are too 'loose'.

(4) *Targets based on work study.* Although work study is based on the parallel techniques of method study and work measurement, it is rare to find method study applied to maintenance. No doubt the main reasons are the nonrepetitive nature of much of the work and the traditional nature of the craft skills employed. However, the first difficulty could be overcome to a certain extent by grouping small items of work and the second by a reappraisal of the skills required for maintenance work.

Having ascertained the facts, the basic questions to ask are:

- What is done? Is it necessary?
- Where is it done? Need it be done at that place?
- When is it done? Need it be done at that time?
- Who does it? Could it be done better by someone else?
- How is it done? Is there a simpler and more economical way?

Work measurement is very largely limited to large direct labour forces employed by local authorities. The building up of a library of standard times would be beyond the capacity of a small firm and would involve excessively high administrative costs. A national library of output values would be of some assistance, but differences from job to job and from region to region are such that the times would require adjustment for local conditions.

Work measurement consists essentially of breaking down a job into its constituent elements of work and repeatedly timing the execution of these elements. An element should be a discrete operation or task which can be easily identified and which has a definite beginning and end. As the element is being timed the operative's speed of working is assessed or rated in relation to a predetermined standard known as 'standard performance', which is represented by 100 on the British Standard Rating Scale. This is defined as the performance that a worker of representative skill and experience can achieve without over-exertion as an average over the working day when motivated by an appropriate incentive. On this scale, 'normal performance' is represented by 75 and is the performance achieved in the absence of an incentive.

The observed times are adjusted to give the corresponding times for a rate of working of 100 *P* and the result is called the 'basic time'.

Basic time = Observed time × (Rated speed of working / 100)

Finally, allowances for personal needs and for working conditions are added to the basic time to give the 'inclusive standard time'. These allowances are of necessity subjective and may form more than 50% of the standard time (see Fig. 9.8). Provision must also be made for time spent in travelling to and from the job, and for preparatory work and clearing up afterwards. As this is related to the size and location of the job, it is better dealt with on a job basis rather than by inclusion in the elemental standard times.

Where the operative receives less than the full value of time saved on the target time, this is known as a 'geared incentive'. For most construction work a 50% gearing is adopted, in which the operative receives half the value of the time saved. The operative starts earning bonus at 50 performance, reaches 33% of the basic rate at 100 performance and continues to rise proportionately thereafter, as illustrated in Fig. 9.9.

In the 50% geared system the target is calculated by adding two thirds to the standard hours. Thus, assuming a work content of 36 standard hours, the target would be 36 + 24 = 60 hours. If the operative were to achieve a performance of 100 *P* and complete in 36 hours, the saving would be 24 hours, of which the

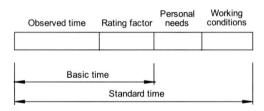

Fig. 9.8   Calculation of the inclusive standard time.

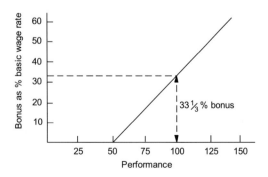

Fig. 9.9   Geared incentive: bonus as a percentage of basic wage rate = $\frac{2}{3}(P - 50)$.

operative would receive half the value (12 hours) representing a bonus of one-third of his or her basic pay.

In view of the uncertainties inherent in predicting target times for maintenance work, a gearing similar to that illustrated in Fig. 9.10 is sometimes adopted. Bonus payments start at 50% standard performance and rise rapidly to 25% basic pay at standard performance and then at a diminishing rate thereafter. This arrangement softens the effect of gross discrepancies in fixing targets and discourages workers from skimping work so as to earn excessively high bonuses.

Indirect incentives

(1) *Profit-sharing.* This may take the form of:
- Immediate profit-sharing, in which the gross profit is calculated on the completion of each job and a percentage of the profit is shared out among the operatives according to the time spent on the job.
- Delayed profit-sharing, in which the overall profits are calculated six-monthly, annually or over some longer period and a proportion is shared out according to some formula. Although immediate profit-sharing provides a greater incentive, delayed profit-sharing is probably more useful in retaining skilled operatives by involving them in the long-term profitability of the firm.

(2) *Merit rating.* This usually amounts to a selective form of 'plus' rates in which additional payments above the basic rates are made for defined factors such as level of output, quality of workmanship, length of service, etc. The extra payment tends to become a permanent addition to the operative's wage and would be difficult to reduce if for any reason the operative's performance deteriorated. It is unlikely that merit rating has any great effect on output but it does provide a means of rewarding the more competent and conscientious workmen, especially where employed on high-quality work for which speed of working would be an inappropriate measure.

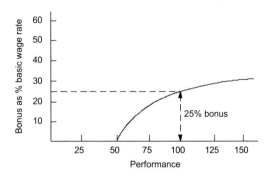

Fig. 9.10 Alternative geared system: bonus as a percentage of basic wage rate $= 50 - (2500/P)$.

The following is a possible method of grading payments:

Skill	Technical knowledge		10	
	Versatility		10	
	Quality of work		10	
	Speed of working		10	
Reliability	Timekeeping		10	
	Responsibility		10	
	Initiative		10	
	Co-operation		10	
Length of service	0–½ year	5		
	½–1 year	10		
	1–5 years	15		
	over 5 years	20	20	
			100	
Grade				
I	0–25*			
II	25–50*			
III	50–75*			
IV	75–100*			

*Additional payments related to comparable rates paid by other firms in the district.

The advantage over the blanket 'plus' rate is that additional payments are related to identifiable qualities, and, over the work-study-based scheme, that speed of working is not the sole criterion.

(3) *Percentage of turnover.* This is confined to supervisors to encourage them to organise and control their workforce to greater effect. It is an attempt to resolve the dilemma created by incentive schemes whereby if supervisors are included in the scheme they have an incentive to pass bad work, while if they are excluded they may well be earning less than the people under their control.

(4) *'Plus' rates.* These are indiscriminate additional payments in excess of the basic rates laid down by the Working Rule Agreement, paid to all operatives merely for the purpose of recruiting and retaining an adequate labour force. They are not related in any way to output and clearly do not provide an incentive to improve productivity.

The above section has demonstrated the importance and application of skills mapping and motivation incentives for maintenance operatives in delivering the maintenance and repair service.

References

(1) Mundell, R. (1996) Repairs by Appointment: The Next Generation. *Institute of Maintenance and Building Management Journal*, Vol. 2 No. 3, Summer, IMBM, Farnham.

(2) Bruhns, H. & Isaacs, N. (1991) The Quality Assessment of Office Buildings. Paper at *Symposium on Management, Quality and Economics in Housing and Other Building Sectors*, Lisbon, 1991.

(3) Barrett, P. (1995) *Facilities Management – Towards Best Practice.* Blackwell Science, Oxford. Ch. 2.

(4) Holmes, R. (1983) Feedback on housing maintenance. *BMCIS Occasional Paper*, July.

(5) Building Research Establishment (1966) *Building Operatives' Work.* Vols. I & II. HMSO, London.

(6) Joseph Rowntree Foundation (1995) Improving the Efficiency of the Housing Repair and Maintenance Industry. *Housing Research Paper 183*.

(7) Association of Direct Labour Organisations (1996) *Building Trade Multi-skilling.*

(8) McCarthy, K. (1997) Multiskilling in Building Maintenance. *Journal of the Institute of Maintenance and Building Management* Vol. 3 No. 3, Summer, IMBM, Farnham.

(9) Advisory Service for the Building Industry (1969) *Incentive schemes applied to building maintenance of small firms.* Department of the Environment. R&D Paper.

Chapter 10
Repair Diagnosis

10.1 Monitoring building deterioration

Condition surveys or repairs inspections carried out for the purposes of building maintenance require accurate identification of the repairs needed. An inaccurate diagnosis can lead to misdirected or inappropriate repair works, or to genuine repair needs being overlooked or sidelined. This in turn may cause greater deterioration and repair costs at a later date, or a breakdown of function affecting the users of the building. These costs may be high in relation to the cost of carrying out the original survey. For example, a surveyor who because of inexperience tends to err on the safe side, may recommend replacing a door when a repair would suffice. This, repeated over tens or hundreds of units, could lead to a repair bill many thousands of pounds higher than necessary; nor would this error be picked up unless a resurvey by a more experienced surveyor brought it to light, which only rarely happens. It is therefore important for the maintenance manager to invest in good survey and inspection practice using suitably qualified and experienced personnel to avoid this multiplier effect.

However, ensuring accuracy in repairs identification is not a simple problem, as there are many variables affecting both the process of deterioration and the point at which this deterioration becomes unacceptable to the users of the building. In general terms these variables are:

- the original specification and installation standard of an element
- its exposure to weather
- its wear and tear through usage
- the performance standards and criteria required by the users.

The performance of a particular building element in a unique location is therefore very difficult to assess and predict with any degree of accuracy on a single brief visit, yet this is precisely how most repair diagnoses are made in the context of the cost and time parameters governing the operation of the maintenance organisation. For example, on a standard quinquennial condition survey, many thousands of separate elements and components will need to be assessed, usually within the space of a few hours. Unless there is a particularly intractable or expensive repair requirement, it is simply not feasible to spend more than a limited amount of time assessing any specific repair need on such a survey. Within this limited time resource therefore, the problem is how to identify repair needs within acceptable limits of accuracy. What these acceptable limits are will depend on the nature of the building, the importance of the element to the building's use, and on the maintenance policy of the users.

An additional complication arises because it is often not feasible for reasons of time and cost to carry out exposure works during an initial repair diagnosis survey. The surveyor therefore must rely on visible evidence only in making an assessment, though this can be supplemented by the use of easily deployed equipment such as damp meters. Even with such equipment, the visible evidence may or may not give a sufficient indication of underlying wants of repair.

There is a further problem, which often goes unrecognised in practice, relating to the subjectivity of an individual surveyor in making repair diagnoses. Studies in the Netherlands[1] and elsewhere have shown that independent surveys of the same defect can result in markedly different diagnoses and proposed remedies, even when the surveyors are from the same organisation. Wide variations may also occur if consultant surveyors are employed depending on how well they are acquainted with the particular operational requirements and future needs of the client's buildings. Further variations occur in the quality and consistency of survey information according to the level of qualification and training of the individual surveyors involved, as indicated above. These variations in standards of diagnosis can be resolved by the maintenance manager to within an acceptable margin by ensuring:

- a standardised system of element recording and condition analysis, appropriate to the nature of the estate[2] (see Appendix)
- accurate briefing of surveyors in terms of identifying when repairs will need to be carried out in the context of the building's use and the maintenance policy of the client organisation
- where possible, access to building records of as-built specifications (note that the CDM regulations now require the preparation and retention of health and safety information relating to maintenance, for relevant new works)
- where possible, access to previous survey records (which by showing element condition some years previously will help to track the rate of deterioration since the last survey).

Moving from these general management issues to the specific, it is now necessary to consider the causes of building deterioration, and the basis on which repairs are identified and recommended.

10.2 Causes of building deterioration

The second law of thermodynamics states that entropy occurs in all physical states and systems; in other words that all systems tend to become more and more disorderly over time. In relation to the manufactured components that comprise buildings and their services, this means that degradation from the original standard of manufacture is inevitable over time. The time span may be short or long depending on a number of factors:

- the quality of the materials and manufacture of the element
- the appropriateness of the element to the environment in which it is installed
- the processes of degradation operating on the element
- degrees of intervention to arrest deterioration (e.g. painting timber)
- the thresholds of performance acceptability to the building users.

It is important to recognise a difference between elements that fail as a result of defect, and those that fail as a result of normal wear and tear. All elements have a design lifespan, over which they can be expected to deteriorate, and to therefore reach the end of their design lifespan. This is itself not a fixed term, but may vary according to the nature of the product and its exposure and use conditions. For example, a lightbulb with a design lifespan of 1000 use-hours will rarely fail precisely at this threshold, but a normal distribution of failure rates for a population of such bulbs will show a strong tendency to group about this median value, even though some will fail sooner, others later. What should be noted is that failure can be expected at around this median value.

On the other hand, a defect can be defined as an unexpected early failure because of faulty design, manufacture, installation, or maintenance. The reasons for this type of failure are outlined in more detail below, but at this stage it is important to note because it may significantly influence our proposed repair remedy for the failure, and may alert us to problems which may be latent elsewhere. Problems with the premature deterioration of concrete using high-alumina cement, or with volatile aggregates reacting with the cement in concrete, are instances of materials defects. Because of these defects, the materials failed before their time, and therefore any repair solution would involve their substitution with more appropriate materials rather than their renewal. On the other hand, elements that have fulfilled their design lifespan at failure and have thus given good service may be replaced with the same or similar products.

In greater detail, the causes of deterioration can be defined as follows:

(1) *Inadequate brief.* It is often said that defects start on the drawing board but in some cases they can originate at an even earlier stage. For example, the brief may lay down totally unrealistic cost limits or fail to give vital information on the functional requirements of the building. Usually there is

no indication of the likely period of use nor of the client's attitude towards maintenance.

(2) *Faulty design decisions.* The most common faults may be grouped as follows:

- Failure to follow well established design criteria in the choice of structural system and selection of materials.
- Ignorance of the basic physical properties of materials, e.g. failing to make allowance for the differing thermal and moisture movements of materials used in combination.
- Use of new materials or innovative forms of construction which have not been properly tested in use. This is often the result of uncritical reliance on manufacturers' literature quoting simulated laboratory tests.
- Misjudgement of user and climatic conditions under which the material will have to perform.
- Complex details that have a low probability of successful execution on an open building site.
- Poor communications between different members of the design and construction teams.

The fault may be traceable to component manufacturers, specialist subcontractors and consultants as well as the main designer.

A less obvious design fault is the failure to consider the ease with which components can be maintained and eventually replaced. For example, little thought is given to the standardisation of components in order to reduce the need to carry a large variety of spare parts or to ensuring that access can be easily gained for servicing and cleaning.

(3) *Construction methods.* The conditions under which construction takes place are often far from ideal and, coupled with an emphasis on speedy completion, can result in careless and skimped work. Although a BRE study[3] showed that only a small proportion of defects were attributable to faulty materials it is apparent that some manufacturers of so-called high technology components have little awareness of the rigours of a building site or the standards of accuracy achievable under such conditions. Thus, whilst the materials may be perfect on leaving the factory they can quite easily be damaged in transit, loading and unloading, unsuitable conditions of storage on site and hoisting and placing in position. Many such defects could be avoided by ensuring greater care at all stages in the process, proper training of operatives, and closer supervision. To tackle this problem the construction industry is beginning to introduce quality assurance (QA) standards such as BS EN ISO 9002. Essentially these techniques consist of setting down appropriate inspection procedures and specifying levels of acceptance and rejection together with methods of sampling and testing the performance characteristics.

(4) *User activities.* Defects may be caused by unintentional misuse through a lack of information on the correct mode of use, or by deliberate acts of

vandalism. The solution is to provide the designer with more information on the degree of severity of use so that a better match can be made between the robustness of the fittings and finishings and the conditions of use. Also, certain defects may be related to the social attitudes and financial circumstances of the user, e.g. condensation is affected by the amount of money spent on heating and ventilating, and the occupancy pattern.

(5) *Maintenance.* Incorrect identification of the true cause of a defect, and inappropriate remedial work, will not only do nothing to rectify the original defect but may substantially worsen the condition of the building. Similarly, lack of care in carrying out repairs and inspections may be the cause of defects in previously satisfactory elements, e.g. walking on unprotected felted flat roofs can drive the gravel into the felt, causing splits and cuts leading to premature leaking. The life of building elements and components can be extended considerably by adopting a planned maintenance approach so that problems can be identified in their early stages and preventive maintenance carried out to avoid early failure.

(6) *Fair wear and tear.* Even with the best quality products properly installed, used, and maintained, failure occurs sooner or later (the time period may be two or three years for, say, a heavily used floor covering, or hundreds of years for stonemasonry). These processes of deterioration, both natural and artificial, are illustrated in Fig. 10.1. An understanding of how they operate is essential to the correct analysis and diagnosis of repair needs and should therefore be an important component of maintenance surveyors' education and training.

10.3 Evidence of wants of repair

Following on from a look at the causes, the evidence of wants of repair will now be examined. Broadly, there are two classes of symptoms: distortion, or a change of shape; and degradation, a change of physical properties[4].

Distortions

Distortions, or changes of shape, may occur as a result of excess loading, both in terms of applied weight, or in some cases, rapid changes in loading, for example wind loads during a storm. It manifests as bending in joists and lintels, warping and buckling of sheet and board materials, or for some materials such as lead. Plastics or asphalt, creep. This process may be irreversible, for example, in creep, but reversible in the case of an undersized joist causing 'springing' in floors. Whilst in some instances a small degree of load distortion may not adversely affect the performance of the component or element, in other cases distortion will affect drainage and water runoff characteristics, seals and junctions, cause cracking in other components (such as cracking in ceilings because

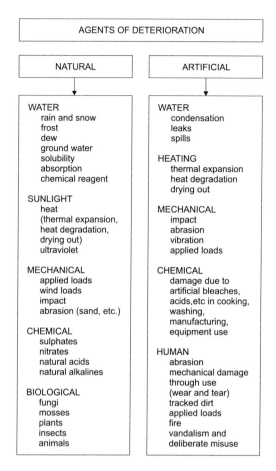

Fig. 10.1 Building materials: agents of deterioration.

of joist distortions, or may alter the structural geometry of the component so that it is no longer structurally stable, for example a bulging masonry wall. The second agent of distortion is change in temperature causing materials to expand and contract, which may trigger serious cracking or buckling, such as in thermal expansion of brickwork or concrete slabs built without adequate movement joints. In the extreme case of a fire, this temperature distortion can be large enough to cause structural collapse, for example of iron or steel columns and beams. Other shape distortions may be triggered by wetting (for example, strawboard flat roofing decks) or drying (e.g. timber shrinkage) or a combination of both, as in subsidence on clay soils; and once again may or may not be reversible.

When identifying and specifying repairs caused by distortion, it is important to understand the causes and specify accordingly, to prevent the repair or replacement suffering the same mode of failure. This will involve looking to minimise the causes of the distortion, or if this is not practicable, specifying a more robust and resistant repair or replacement.

Degradations

There are many agents of degradation, or the change of physical properties of a component, which may act on a component either singly, or more usually in combination. The chief physical agents are water, heat, ultraviolet, and chemical reactions. Biological agents include insects, fungi, mosses, plants, animals, and humans. Mechanical agents comprise impacts and vibrations. Fire may also be considered an agent in its own right.

The way in which the physical and some of the biological agents act is by causing or triggering chemical reactions in the material which transmutes it into a less appropriate material, for example iron to iron oxide in the rusting process. Other examples may include the bleaching effect of ultraviolet light in sunlight causing loss of colour, and in the case of plastics, a breakdown in the molecular chain structure leading to brittleness. Water, either on its own or in combination with other agents, is a major cause of degradation, acting in a number of different ways. Water can degrade by solubility (plaster and limestone, for example), absorption (leading to swelling), change of volume (freeze/thaw cycle), mechanical (abrasive effect of running water), and in the formation of salts and crystalline structures (hygroscopic or water-attracting nature of sulphate salts in rising damp). In addition or in combination with water, natural and artificial chemicals in the environment, such as sulphates in soil or nitrous oxide from vehicle exhausts dissolved in rainwater to form a weak acid (acid rain) may react with a wide range of materials such as concrete, plaster, stone, and exposed metal to cause deterioration. The presence of water may facilitate attack by biological agents, particularly wood-rotting fungi and plants.

Biological agencies may devour the material (wood-rotting fungi, mosses, insects, rats), or stress the material mechanically by physical wear and traffic (plant roots forcing components apart, abrasion and damage caused by larger animals and humans). Mechanical agencies such as vibration and impact cause physical damage because of shattering, consolidation (such as loose fill materials subject to vibration), or stressing to breaking point (impacts and loads). Dirt and debris can also be considered agents of degradation, attacking materials by deposition (loss of finish), or by abrasion (dirt in moving parts such as pumps and motors; scouring effect on floor coverings), or by clogging gaps and pores, for example dirt in carpets.

As with repair identification for distortions, those involving degrading agents need an analysis of the causes, and whether these causes can or should be lessened as an integral part of the repair. For example, it has been found that a simple remedy such as placing an effective entrance doormat which will remove incoming particles from the soles of feet, can double the life of floor coverings within a building. If the causes cannot be lessened, the repair solution may involve the use of higher-specification materials that will prove more resistant, provided there is no undue cost penalty for so doing.

Lastly, fire may also be considered an agent; an event rather than a process, but one that is very destructive owing to combustion, heat damage, smoke damage. Removing the causes of fire and ameliorating its effect by the use of fire-resisting construction, installation of automatic alarms and defence such as sprinklers, is a large and well-documented subject area, usually subject to statutory control (cf. Chapter 3).

These agents work either singly, or more commonly, in combinations on all building and service elements. Further variables include the intensity and the timescale over which these agents operate. It follows that for each particular component in a building, the pattern of degradation will be unique. For example, similar windows will deteriorate differently depending on which face of the building they are on and thus their level of exposure. Those on the north face may be more prone to moisture and biological attack; those on the south, from UV sunlight. They will therefore deteriorate in a different manner and over different timespans.

10.4 The concept of performance in repair diagnosis

These processes and effects of deterioration and deformation may or may not affect the performance of a building element adversely. For a properly specified and installed component, the operation of these processes over its design lifespan should not lead to a failure in its performance. In other words the component will continue to function to an acceptable level even whilst it is deteriorating. Furthermore, the rate of deterioration will depend on the degree of intervention (if any) by the maintenance operation. Regular cleaning and painting operations will tend to prolong the life of an element by holding up or even reversing the causes of deterioration. External timber work is perhaps the clearest example of this, though it is also very significant in relation to service installations such as heating, air conditioning, and lifts, where regular lubrication and adjustments delay the processes of wear and minimise the likelihood of sudden failure. Failing to assess the effect any such interventions may have on remaining lifespans of building components is a common shortcoming in condition survey practice. However, there comes a point where the component is no longer fit for its purpose. Determining exactly when this point has come is at the heart of effective repair diagnosis, and this is why the concept of performance is crucial (see Fig. 10.2).

In broad terms, the performance of a component may be defined as its ability to satisfy the requirements of the user, and therefore any deterioration that does not or will not affect this ability can be disregarded. However, assessing levels of deterioration which do not affect current performance but which may do in the near future is important in condition surveys for planned maintenance

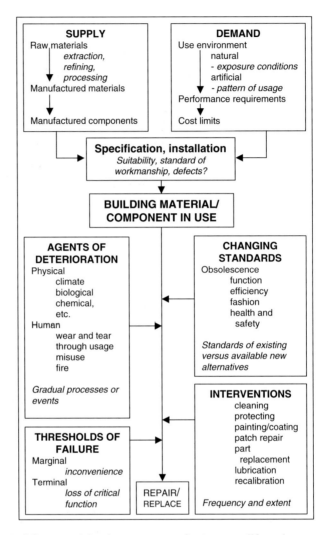

Fig. 10.2   Building material and component performance – a life cycle.

programmes. Performance criteria are many and varied: some, such as structural stability, apply to nearly all components, whilst others such as appearance may affect some but not others. An indicative map of component properties and performance requirements is given in Fig. 10.3.

Indeed, a difficulty in defining any general performance requirements in building maintenance lies in the fact that user requirements for the same component may differ widely, and may also change over time. As an example of the first, a coat of paint on a factory shed door may be required only as a protective coating, but on a shop or house door may have the additional function of improving appearance. It therefore follows that a dirty coat of paint may be satisfactory for one use (the factory), but unsatisfactory for the others. As an example of

---

BUILDING MATERIALS AND COMPONENTS

PROPERTIES	PERFORMANCE
What the material is:	*What the material is*
Appearance	*required to do:*
finish, colour, texture,	Health and safety
transparency/opacity	minimise dangers
Structural form	from physical,
bar, sheet, loose,	chemical, electrical
cast, formed	injury
weight, density	Structural integrity
Structural characteristics	maintain form
strength in	Structural performance
compression/tension,	carry applied loads
stiffness/elasticity	Mechanical performance
Water permeability	reliable operation
wetting/drying	Electrical performance
Chemical properties	insulate; conduct
inertness/volativity,	Thermal performance
photochemical	insulation, or heat
Biological properties	dissipation functions
composition, stability	Acoustic performance
resistance to attack.	passage of sound
effect on humans,	Resistance to water
animals and plants	resist passage of
Thermal characteristics	water; resist
'K' values	damage by water
Acoustic characteristics	Resistance to chemicals
sound transmission	resistance to attack;
properties	reactions with other
Mechanical operation	components and
Electrical operation	chemical agents
insulation,	Resistance to biological
conductance	attack
Behaviour in fire	fungal, insect,
combustibility	animal, moss and
flame surface spread	plant
fire resistance	Resistance to misuse
Maintenance and servicing	accident,
requirements	vandalism, fire
cleaning, coating,	behaviour
lubricating, adjusting	Other component-specific
Cost and availability	requirements

Fig. 10.3   Building material and component properties and performance requirements.

performance requirements changing over time, a lead water pipe, acceptable in the nineteenth or early twentieth century when installed, is now considered to be unsatisfactory because of health risks. The component may still be fulfilling its design function to a satisfactory standard, but because the performance requirements have changed, it has become obsolete and therefore needs to be undertaken as a repair.

Thus whether at the time of a maintenance inspection the performance requirements for a particular component are satisfied will depend not only on the physical and functional state of the component but also on the particular performance requirements of the user at that time. It follows that accurate diagnosis of repair needs can only be made in the context of understanding the user's specific performance requirements for that component. This point refers back to a point made at the beginning of this chapter, the need for adequate training and briefing, particularly of consultant surveyors, prior to repair inspections, in order that diagnoses may be accurately matched to the ongoing needs of the users.

Another issue remains in the definition of performance, that of thresholds. For many building components the threshold as to when it passes from serviceability to unacceptability is not clearly defined. For example, at what point would deteriorating mortar in a brick wall trigger repair action? Certainly if it was so far advanced that the structural integrity of the wall was in question, but before that point is reached, the wall would continue to fulfil its function even whilst the lack of mortar would be accelerating other deterioration processes such as water damage and frost attack to the brick faces. The performance threshold is thus a grey area rather than a well-defined line in many instances, and this can be the root cause in problems of surveyor subjectivity referred to at the beginning of this chapter. In this context it may be prudent to adopt two thresholds for repair action:

Terminal performance failure
When the component ceases to fulfil critical functions such as mechanical function, structural stability, or health and safety requirements. Repair action will be necessary.

Marginal performance failure
When the component's performance may adversely affect the building users (either functionally or aesthetically if relevant, such as outdated but sound decorations), but not critically.

Such a distinction may be represented in the priority assigned to a particular repair (cf. Chapter 4), where some repairs critical to core performance may be done before other more marginal wants of repairs.

10.5 Repair diagnosis

The general approach should be to examine carefully all the symptoms, consider all the probable causes, by a process of elimination identify the true cause and its source, and then decide on appropriate remedial action. In many instances a visual examination will be sufficient for an experienced inspector to determine the cause. Where this is not possible instruments may be used for a more objective diagnosis. There are many portable instruments on the market for investigating the causes of defects, including:

- Moisture meter, to test the comparative moisture content of materials.
- Ultrasonic tester, for checking concrete strength.

- Depth meter, to enable the cover to reinforced concrete to be ascertained. This instrument can also be used to locate the presence of metal pipes in unreinforced concrete.
- Endoscope, which comprises a slender tube fitted with a magnifier and light that can be inserted through a small hole to give a view of the interior of wall cavities and other voids within a building.

In some cases it may be necessary to monitor the rate of development of a defect over a prolonged period in order to assess the probability of failure. Examples include measuring the width of cracks by means of pins and a micrometer, taking levels at fixed points to determine the rate of settlement and checking vertical alignments with the aid of a theodolite.

Where the surface manifestation of a defect is not sufficient to indicate the true cause, and particularly where there is a reasonable possibility that a structural defect is concealed, an exploratory survey should be undertaken. Such investigations should be carried out carefully to minimise consequential damage, and can be facilitated if the design allows ready access to hidden parts of the structure, e.g. strategically placed access panels to ducts and ceiling hatches. Specialist advice may be required in respect of certain defects, e.g. structural stability and services and laboratory tests of subsoil or materials to establish their composition and performance characteristics.

For more detailed information the reader should refer to the relevant BRE digests and to other published sources listed in Further Reading and Contacts.

References

(1) Damen, A. (1992) Condition-based Building Maintenance. Keynote speech at *Chartered Institute of Building International Conference: Management Maintenance and Modernisation of Buildings*, Rotterdam, 1992.

(2) Wordsworth, P.(1992) *Standard Maintenance Descriptions. Research Paper 20,* Royal Institution of Chartered Surveyors, London.

(3) BRE Digest 176 (1976) *Failure Patterns and Implications.* HMSO, London.

(4) Everett, A. (1986) *Mitchell's Building Series: Materials,* Mitchells, London.

Chapter 11
Conservation and the Environment

11.1 The maintenance manager as conservationist

Since the 1970s there has been a widespread movement in western society towards an appreciation of damage that may be done, both culturally and physically, in the name of unchecked progress in science and technology. This has affected many aspects of technology, and many of these issues are of direct concern to built environment professionals in general and maintenance managers in particular. The reasons for these concerns are historical. Prior to the twentieth century, the impact of humans on the existing built environment, the ecosystem and the weather was limited by the then-current levels of technology, whereas especially in the latter part of the twentieth century, the potential and actual capacity of technology to radically change these fundamental environments became much greater. Added to this is the growth in populations, better living standards, and the growth of a global commercial culture which, unchecked, tends to value short-term efficiencies and profit maximisation over other less tangible measures of progress such as cultural identity and environmental quality. In the built environment this was typified by the now-reviled modernist movement in architecture, which in its latter degenerative phase produced many buildings out of scale and out of context with the existing built environments. Such buildings were expensive and difficult to maintain because of the nature of the materials and forms of construction used. Furthermore the buildings were unpopular with a large section of the general public because of their soulless nature. The short-termism of much development during the latter part of the twentieth century, when buildings were often designed with a lifespan expectancy of as little as twenty years, added to the public's unease towards these trends, encapsulated by Prince Charles in 1982 [1] in his unflattering assessment of much of modern architecture.

Whilst some of this opposition to science and technology has become extreme in its rejection of any and all new building or infrastructure projects, there is now in place a consensus which, whilst recognising the substantial benefits that progress in these areas brings, nevertheless wishes that its impact on the existing environment, both physical and cultural, is assessed with a view to limiting or halting potentially adverse effects. In terms of the built environment this has led to the establishment of an influential conservation movement which seeks to preserve the best of our older buildings and environments, and an environmental movement which seeks to encourage the sourcing of building materials from sustainable sources, to control the generation of pollutants such as carbon dioxide by buildings in use, and to limit adverse health risks in buildings.

The growth of the environmental and conservation concerns has paralleled the increasing emphasis on maintenance and refurbishment of buildings as an alternative to redevelopment in recent years. Indeed it can be seen from the above arguments that good maintenance management actively promotes many of the aims of conservationists and environmentalists in terms of maximising a building's effective lifespan, minimising energy and materials resource consumption, and in preserving old buildings by keeping them fit for modern use as an alternative to redevelopment. These 'green' maintenance management credentials deserve and demand better promotion and recognition, to enhance the status of the maintenance management profession in comparison with that of the comparatively environmentally wasteful but higher profile architecture and construction professions.

For practising maintenance managers, increasing emphasis on conservation and environmental issues has highlighted the need for ongoing professional development to keep abreast of changing legislation (Chapter 3), and new methodologies in the fields of preservation, energy management, waste management, and environmental impact assessments. An overview of these areas is therefore given here.

11.2 Conservation

Conservation can be defined as 'all the processes of looking after a place so as to retain its cultural significance. It includes maintenance, and may according to circumstance include preservation, restoration, reconstruction and adaptation and will be commonly a combination of more than one of these.' (ICOMOS, Burra Charter, 1981.)[2]

In general terms, conservation principles and concepts are applicable to any building which has features of historical, architectural, or general cultural interest worthy of preservation. In practice, a manager with little knowledge of architectural and cultural history generally, or of a particular building's history, may find it difficult to recognise such features. However, there are a number of statutory and voluntary bodies and interest groups who now act to identify and protect such buildings. The most notable of these are: English Heritage, the government agency responsible for historic building legislation standards (Chapter 3), and itself the owner of about 600 historic buildings and monuments; the National Trust, a charitable body that owns or manages many hundreds of historic buildings and conservation sites; and the Society for the Protection of Ancient Buildings (SPAB), an influential conservation pressure group founded in the nineteenth century by William Morris and others. There are other special interest groups including the Georgian Society and the Twentieth Century Society who are concerned with the conservation of specific classes of buildings.

The movement to conserve worthy buildings and monuments coincided in Britain with the massive expansion of the built environment triggered by the industrial revolution in Victorian times. The writer and architectural critic John Ruskin, considered by many to be the father of the modern conservation movement, put forward the view that an increasingly technologically powerful and market-oriented society was in danger of obliterating its architectural heritage in the name of progress; and along with it, its collective cultural memory and sense of history. This heritage vandalism was being caused both by the wholesale destruction of worthy old buildings; and, even worse according to Ruskin, by inappropriate, sentimentalised 'restoration' which obscured the truth and historical worth of a building and replaced it with a shallow pretence of historical accuracy: – 'scraped and patched up into smugness and smoothness more tragic than uttermost ruin' as Ruskin himself puts it[3].

Many maintenance managers committed to the concept of good building maintenance management practice as enhancing the physical comfort and aesthetic appreciation of our buildings will probably empathise with the broad goals of conservation whilst sometimes being baffled or dismayed at the ways in which too 'precious' an approach to conservation issues may interfere with the safe and economic running of a building.

There may be serious conflicts relating to obsolescence in function, space usage and services, health and safety standards, fire defence, between the historic fabric and services in a building and the required modern standards of use. Other problems may arise relating to the repair or necessary replacement of historic elements at the end of their functional life through fair wear and tear. Not everything that was ever built can or indeed should be preserved, and the requirements of conservation may often impose an unwelcome economic or functional cost on the present use and efficiency of an historic building. This ambiguity is at the heart of the relationship between building conservation and building maintenance management, and to resolve it, there is a need to have a framework for decision making such as the following:

- the specific reasons for conserving all or part of a building or monument (conservation philosophy)
- how and in what ways this will then affect its value and utility (concurrence and conflicts of interest)
- how we may evaluate the conflicts and benefits of the conservation process on the use and management of the building as a whole (a framework for evaluating conservation), and
- by what means we carry out the conservation (sourcing skills and expertise).

11.3 Conservation philosophy: concurrence and conflicts of interest

There are many differences of opinion in what conservation is or should entail. At one extreme there is the antiquarian or art historian's view that the artistic and historical worth of an historic building should transcend mere short-term economic or functional considerations, typified by early conservationist William Stukeley's 1751 condemnation of alterations and demolitions to such buildings as being done merely 'for the dirty little profit'. Ruskin and Morris themselves were inclined to this purist view, though it should be noted that much of Morris' own factories' output of stained glass and wallpaper was destined for renovation schemes on many historic buildings which in fact destroyed much of their historic fabric. At the other extreme we have those who think conservation should pay its way in the economics of the property marketplace, and if it cannot, then the building should be sacrificed; for example the art deco masterpiece the Firestone building in London was demolished for redevelopment the day before a spot-listing was about to come into force.

In between these poles lies the broad consensus that historic buildings need to be conserved, though with due regard to the economics and functionality issues that this raises. This however demands a clear definition of conservation and how this relates to the idea of restoration; that is, restoring functionality, for example, to the weather envelope to prevent decay; or to remodelling the spaces and functions within an historic building. Clearly, one problem with historic buildings is that their original function and purpose may be long obsolete, and thus some degree of alteration for a new use is inevitable, even if this means no use. For example, no castle or fort now retains its original defensive function; and many churches and cathedrals, whilst still in use for religious purposes, no longer fill the community functions which led to marketplaces in the naves and weddings in the porches in their heyday. Indeed, there is one rather extreme conservation school known as the Decadents who maintain that once the original use is abandoned the building is best left to decay naturally and with no repairs or other interventions to prolong its life artificially, because so doing would destroy its remaining integrity and worth forever. Needless to say, this philosophy of romantically wasting assets holds little currency today.

On the other hand, the restoration school of conservation seeks to put back the clock and restore the historic building to the state it was when at its historical or functional peak. The problem with this approach is, exactly what point in time do you put the clock back to, given the fact that most historic buildings have themselves evolved over the years by being altered, repaired, and added to, and the evidence of this evolutionary process may itself be part of the building's historical value. For example, many mediaeval cathedrals took hundreds of years to build, and are in fact an accretion of different styles and ages; a process which continues even now, as new stained glass, new stonework, and new heating and electric services are installed in what remain essentially working churches. In fact, early restorers such as the subsequently vilified James Wyatt, and the

Cambridge Camden Society, in the early nineteenth century decided that the purest period of English ecclesiastical architecture was the decorated gothic period of the early fourteenth century, and hence their 'restorations' would consist of stripping away any subsequent alterations and additions, however historically worthy, to be replaced with modern copies of what they presumed the original fabric must have been. There are strong echoes of this today, more commonly in domestic architecture, where the 'reinstatement' of timber beams or inglenook fireplaces which were never there in the first place remains a corrupting fiction of conservation, despite the obvious comfort that such 'period' details may bring the occupiers.

11.4 A framework for evaluation

So where, then, can maintenance managers turn for guidelines in the ongoing conservation philosophy debate if they need to deal with an historic building? In view of the legislation in force on many of our historic buildings, the first resort should be English Heritage, the government body responsible for advising on listed building legislation and which manages 600 of the country's historic buildings and monuments. Its document *The Principles of Repair*[4] gives the following guidance:

> 'The authenticity of an historic building … depends crucially on its design and on the integrity of its fabric. The unnecessary replacement of historic fabric, no matter how carefully the work is carried out, will have an adverse effect and seriously reduce its value as a source of historical information.'

Therefore, as the first principle, old fabric should be covered up, not stripped out. Preservation wherever possible, even of damaged or decaying items, is to be preferred. If they no longer meet a critical function such as structural stability or weather tightness, then parallel modern solutions could be introduced to supplement or replace their function. These 'modern' solutions should of course wherever possible be of materials and using skills in sympathy with the original.

The second principle concerns how necessary repairs and alterations should be done. Any additions to an historic building should look what they are: modern additions, and should never pretend to be original by being artificially weathered or distressed, nor by being of a material which pretends to be another, for example, plastic as wood: that is a deceit. In practice this will entail sourcing appropriate materials and skills so that the repairs and alterations complement, rather than contrast with, the original wherever possible. In this context the 'two metre rule' can be profitably applied: to keep the overall appearance and style of a historic building or feature, repairs and alterations can be seen to blend in with the original, to present a harmonious whole, from a distance of greater than two metres. Within that, there should be a subtle but honest distinction between the original and the new materials. Such sensitive restorations and additions may thus become themselves worthy of conservation in the years to come, or if not, they can be painlessly removed. The writer Stewart Brand puts this point succinctly in his book *How Buildings Learn* (1994): 'as much as possible of the original fabric is to be saved. New work should be potentially reversible[5].'

The third principle of conservation is one well known to maintenance managers: avoid the need for restoration as far as possible by good maintenance practice throughout the life of the building. As William Morris puts it in the manifesto for SPAB in 1877: 'put protection in place of restoration, stave off decay by daily care, and show no pretence of other art[6].' Ruskin again expresses what has continued to be true: 'the principle of modern times ... is to neglect buildings first, and restore them. Take proper care of your monuments and you will not need to restore them[7].' The repairs thus carried out should be appropriate to the nature of the building, using materials which complement, (not necessarily which match) the original fabric – what Ruskin calls 'an honest repair'.

There remains the significant problem of those alterations deemed necessary to modern health, safety and comfort but which may nevertheless cause damage to the historic structure. Environmental enhancements such as heating may require especial attention, as an unwelcome consequence of heating an historic building could be drying out of plaster, causing cracking, or of timber, causing shrinkage and distortion. It may be necessary therefore to install humidity controls to counter such effects. Fortunately, as technology advances, it becomes more feasible to introduce such intelligent, tailored responses to these environmental problems caused by upgrading to modern standards.

Other similar problem areas include access for people with disabilities, fire defence requirements, and health and safety issues linked, for example, to the presence of lead paint or other potential toxins in the historic fabric. Whilst it is beyond the scope of this book to offer detailed technical solutions to this considerable raft of potential problems, a sensitivity to the issue is the essential prerequisite to a successful solution which permits modern usage with minimal damage to the historic fabric. However, all buildings in use generate some wear and tear, so some unavoidable degree of degradation is perhaps not only inevitable, but, properly managed, may actually be seen to continue the progress of an historic building's evolution rather than to artificially preserve it. It is the honesty and appropriateness of the ongoing preventive and repair regimes in place to manage such wear and tear that matters in such cases. This concept of 'evolutionary conservation' deserves more discussion and development in the context of maintenance management.

Having arrived at these broad principles of conservation, it is necessary to consider how they may be applied to a particular building. In most cases the conservation legislation will dictate what is to be conserved according to the listed building system referred to in Chapter 3:

- *Grade I listed:* 6000 UK properties of major historic interest. This group includes monuments such as Stonehenge, and many major cathedrals. Any repairs and alterations are very strictly controlled.
- *Grade II* listed:* 23 000 properties of significant historic interest, including many country and town houses.
- *Grade III listed:* 400 000 properties of special interest, that is, with some particular notable feature, or part of a conservation area.

This system is broadly comparable with the conservation 'hierarchy' of buildings:

(1) The 'don't touch at all if possible' class of monuments such as Stonehenge, and significant 'jewels' of buildings, where any alteration or addition would detract from its historic and artistic worth. There may nevertheless still be conflicts with fire safety in particular, and health and safety requirements generally, if the building is to be used, visited or occupied. Very sensitive handling of any repairs and alterations is needed, which should entail consultations with conservation groups such as the Georgian Society as well as with English Heritage and the local authority's conservation planning officer.

(2) The 'handle with care' buildings, which may acquire a modern use (notably tourism such as the majority of National Trust properties) which require additions and adaptations. Historic elements should remain undisturbed wherever possible, repairs should be distinguishable from the old but in sympathy with their materials and style, and new installations should be replaceable with minimal damage to the historic fabric.

(3) The 'feature' buildings where it is a particular feature of the building rather than its whole fabric which is important. This group includes those buildings in conservation areas where a certain external appearance must be kept (for example, no concrete tiles or uPVC windows permitted); 'blue plaque' buildings where someone famous once lived (everyone from Charles Darwin to Jimi Hendrix), or buildings with a particular feature of note: an Adams fireplace, distinctive wood panelling, and so forth.

Specific conservation needs and legal requirements can best be interpreted within this framework.

11.5 Sourcing skills and expertise

Deciding what needs to be done to repair or alter an historic building is one thing; putting this into practice may be quite another. The first problem will be obtaining the agreement of the local conservation officers to the works. This will require:

- The preparation of an accurate and comprehensive record of what is there before any works start. English Heritage publish guidance on this[8].
- Careful preparation and specification of works proposals.
- Consultation with the local authority conservation officer, the users, owners, and other interested groups, before work starts. This includes obtaining any necessary listed building consents required under the town and country planning statutes (Chapter 3).
- Sourcing of required works and materials.
- Supervision of works in progress, including the recording and imaging of parts of the fabric that have been opened up.
- Subsequent monitoring of the repair to ensure no longer-term adverse consequences are causing damage (for example, new zinc flashings next to existing copper sheeting, which will trigger electrolysis).

The chief problem the maintenance manager will be faced with in this regard is where to find the materials and skills to carry out nonstandard building work. Old skills such as lime plastering, thatching, traditional carpentry, and stonemasonry are still extant but scarce. Thus there is also a cost problem: conservation work will cost more per unit than normal repair or alteration work. However, if the building is listed, there may be grant aid from English Heritage to fund the difference in cost between a 'normal' repair and one using traditional techniques and materials required as a consequence of its statutory listing. European Community cultural grants and in some cases national lottery funds may be available, but usually only for those historic buildings to which the public will have access. Other conservation managers may be able to help with sourcing appropriate builders and craftspeople, though there is an increasing need for a reliable register of such companies.

In conclusion to this section, there is a need for the maintenance manager to be aware of conservation issues in general, and as they relate to a specific historic building under his or her care, as the essential ingredient in ensuring the correct approach to repairs and alterations. A light and honest touch, respecting the story the building has to tell whilst ensuring its continued use and hence preservation, is needed to ensure that the manager's stewardship adds value specifically to the building itself, and more generally to our culture and society.

11.6 Environment

The term 'environment' is used to describe a range of general issues relating to the adverse impact that modern technologies can have on the natural world.

Whilst some extreme groups regard almost any technology as bad for the environment by its very nature, there is an increasingly large consensus that the benefits that technology brings must be balanced against the damage that may be caused to the natural world; and where possible, the potential for such damage should be either eliminated or reduced to an acceptably low level. As the built environment is the major consumer of both materials and energy (over 50% of the total energy used is in connection with building occupancy[9]), it follows that the building maintenance manager has a central if hitherto unrecognised role to play in minimising environmental damage caused by building procurement and use. Broadly, this can be done by extending the life of existing buildings and structures by efficient maintenance, by reducing energy consumption to an acceptable minimum, and by sourcing energy and materials used in the repair and running of a building in an environmentally responsible manner. It should be noted that achieving these environmental goals may sometimes need to be balanced against greater costs incurred, which may tend to make the building uncompetitive in a commercial situation. However, there can be positive benefits from environmental management in the form of reduced costs over a long time span (for example, in energy efficiency), in projecting a more responsible environmental image, and in improving health and safety conditions within the building.

The environmental impact of buildings occurs at several levels:

- *global:* global warming and ozone depletion
- *national:* energy usage, depletion of natural resources, acid rain
- *local:* conservation areas, waste generation, traffic generation, greenbelt encroachment
- *building:* conservation, energy use, sick building syndrome, materials sourcing, nuisance and pollutant output, health and safety relating to building materials.

The most immediate benefits of efficient environmental management will obviously be at the building or local level. Above this, the individual contribution of a single building or property portfolio will be insignificant, and therefore environmental concerns are managed by the imposition of legal standards at both the national and supranational levels. An example of this can be found in Part L of the Building Regulations, which imposes thermal insulation standards on new construction and refurbishment which would not otherwise be met because the economics of the property market would offer only a very low rate of return on such an investment. As the impact of phenomena such as global warming and resource depletion become more serious, further environmental legislation may be anticipated, triggered by international agreements such as the Kyoto Protocol to reduce carbon dioxide emissions, and by national and local measures which seek to impose tax burdens on polluting processes under the 'polluter pays' principle, for example as is currently in force for waste management under the Environmental Protection Act 1990.

The building maintenance manager therefore needs to integrate environmental concerns into the general management framework of building costs and functions for these reasons:

- to comply with statutory requirements relating to the environment
- to ensure resource efficiency under the 'polluter pays' doctrine
- to promote a good public image and to fulfil the social responsibilities of the building owners/users
- to ensure the health and safety of building occupants and neighbours.

Since these concerns are diverse and often unrelated, it is therefore good practice to adopt an environmental policy in building maintenance management, as an integral part of the overall maintenance policy (Chapter 1). This policy may integrate with an organisation's wider environmental policy, which may also cover areas outside the remit of building maintenance such as products, transportation, and the siting of operations. In its publication *Environmental Action Guide for Building and Purchasing Managers* the Department of the Environment (now the DETR) advocates a strategic approach:

> 'Inadequate building maintenance leads to the wastage of natural resources just as much as the wastage of investment resources. The need for "clean" technology in the operation of building services is becoming increasingly insistent. High standards of environmental performance in building management require high standards of building maintenance.'[10]

The guide goes on to give a more detailed agenda of measures to be taken, the most significant of which are:

- The Environmental Policy Statement, which should cover:
  - o procurement and use of building products and materials
  - o standards of servicing and maintenance
  - o good practice in pollution prevention and waste management
  - o operating procedures to ensure energy efficiency and waste minimisation
- Contract Guidelines, to cover:
  - o standards for servicing and maintenance routines
  - o prohibition and limitation of usage of pollutant products such as halons and CFCs
  - o procedures for safe handling and disposal of waste materials
  - o preference for use of recycled materials and products where possible
- Emergency call-out measures to deal promptly with leaks and spills

- Awareness of landlords where they are responsible for maintenance.

11.7 Environmental building audit

The guide advocates an environmental building audit as a first step to examine the scope for improving environmental efficiency, which could coincide with a more general building condition audit carried out as part of a planned maintenance programme. It advocates planned maintenance and servicing routines as being environmentally beneficial in reducing energy use, materials consumption, and preventing environmental disasters such as legionella poisoning. An eighteen point agenda for the audit is given:

(1) standards of internal environmental quality
(2) environmental efficiency of workplace layout and space utilisation
(3) acoustic quality
(4) air circulation and quality
(5) thermal insulation efficiency
(6) operating efficiency and control of air conditioning, ventilation, heating, and electrical systems
(7) energy efficiency and management
(8) lighting efficiency
(9) potential for waste heat recovery and solar gain utilisation
(10) water conservation and quality
(11) operational efficiency of soil and waste systems
(12) risk level of pollutant emissions and internal pollution; and standards of environmental safety
(13) location and safeguarding of potentially pollutant or hazardous building materials
(14) operational efficiency of fire prevention systems
(15) adequacy of waste disposal/recycling facilities
(16) condition of building fabric
(17) scope for external environmental improvement
(18) efficiency of maintenance (and conservation) procedures and achievement of maintenance standards [10].

Such an audit is also a requirement for organisations seeking accreditation under BS 7750: specification for environmental management systems [11], which lists four broad areas for review:

- evaluation and recording of significant environmental defects
- review of legislative and statutory requirements
- review of existing environmental controls and practices
- feedback from any previous pollution incidents.

The scope and nature of the environmental audit will obviously differ according to the nature and use of the buildings or estate, and in context with the wider environmental concerns of the occupier. The initial audit should seek to be a one-off comprehensive review of environmental issues affecting the use of the building. Subsequently, it is necessary to introduce systems for ongoing environmental monitoring to ensure that key processes and operations, ranging from contract procedures to air conditioning plant maintenance, are kept under review on a regular basis. One recognised method of environmental audit that may be undertaken on new projects is by the Building Research Establishment and is known as BREEAM [12].

The above section has highlighted the need for building maintenance managers to adopt a comprehensive assessment method and policy statement to address environmental issues in building maintenance. Within this broad management context there are several specific topics which will be dealt with in greater depth:

- energy management
- building environment
- procurement of materials, goods and services
- pollution and waste management.

11.8 Energy management

Buildings consume virtually half the UK's energy consumption[10]. Most of this is obtained from fossil fuels such as oil or gas, or from nuclear fuels, although a small but increasing proportion comes from hydroelectric or wind-driven turbines. The environmental impact of energy use is manifest primarily in the production of carbon dioxide as a consequence of its generation, which contributes to the 'greenhouse effect' and hence global warming, and also to the production of acid rain, caused by the gases emitted from power stations dissolving in atmospheric water to form weak acid in rainfall which can interfere with the growth of trees. Therefore the environmental goals to reduce this are, firstly, to source the energy in sustainable ways that do not cause pollution, and secondly, to keep the consumption down to the minimum necessary for the building's acceptable functioning. In respect of the former, there is little the maintenance manager can do except to try to maximise the building's potential for solar gain and natural ventilation. Whilst worthwhile, these gains are likely to be slight in comparison with the scope for maximising energy efficiency and reducing consumption.

Energy is used in buildings in a number of ways. The main use is space and water heating, and the amount used for this will correspond with the required comfort levels in the building and the efficiency of both the thermal envelope

and the heat-producing and distributing systems within the building. Other en-
ergy-consuming operations include principally: lighting, mechanical ventilation, and the operation of motors to lifts, escalators, gates and pumps. Computers, cookers and ranges, and other equipment and processing plant within the
building may also account for significant energy consumption depending on the
use of the building. Scope for environmental gains in energy usage will therefore usually arise from making these processes more efficient in their delivery
rather than by restricting or limiting the use of the building such as by reducing
the temperature to a level which is inconvenient or unacceptable to the building
users. It is recognised that building user education has a limited role to play.
Users can be encouraged to switch off lighting when not in use, or to turn down
radiator thermostats rather than opening windows when too hot; but too intru-
sive a regime is likely to either irritate the users or be ignored. The maintenance
manager can better focus on improving the efficiency of the systems which use
energy by such measures as installing 'intelligent' controls such as movement
sensors which turn off lighting if no movement is detected for a certain period, or
by replacing luminaires with more energy efficient models during maintenance
operations.

The Energy Efficiency Office[13] has indicated an average saving on energy
costs of about 20% is achievable through energy conservation measures, with
an average payback period of 18 months on the investment. The payback on the
capital invested in energy saving measures, for example the extra cost of low-
energy luminaires to replace existing ones, may be assessed using the discounting techniques covered in Chapter 5. To implement energy conservation measures, it is first necessary to carry out an energy appraisal, possibly as part of a
wider environmental building audit referred to above.

11.9 Energy appraisal

There are a number of recognised routes by which an energy appraisal may be
carried out, either by the maintenance manager or, at a higher level of detail, by
specialist energy use consultants. A guide useful to maintenance managers was
produced by RICS and the Energy Efficiency Office under the title *Energy Appraisal of Existing Buildings*[14]. The guide gives a systematic checklist for the
appraisal of energy use, gives benchmark energy performances for classes of
buildings for comparison, and has further guidance for more detailed information. The guide proposes three levels of appraisal according to available resources and expertise:

- Level 1: guidance involves simple steps appropriate to novice energy ap-
  praisers
- Level 2: requires more experience and specialist advice should be carefully
  considered

- Level 3: measures allow the surveyor to identify areas, which will almost certainly require specialist consideration.

The process consists of four steps:

(1) pre-inspection: research and collation of relevant existing plans, records and reports
(2) building fabric survey: assessment of existing insulation, glazing, day-lighting, ventilation, areas of condensation, etc.
(3) energy supply survey: assess metering, loadings, CHP efficiencies, fuel storage facilities
(4) services survey: assess space heating, incinerators, hot water, mechanical ventilation, air conditioning, lighting, office equipment and processes, building management system.

The appraisal of the survey findings indicates how savings and efficiencies may be made according to the level of expertise (see above) and available finance, and indicates some investment appraisal methods.

Whether or not a formal energy appraisal is undertaken, it is still possible to effect energy savings during the routine management of maintenance by having regard to the issue when specifying repairs, replacements and servicing routines. For example, routine annual inspections and servicing of plant afford the opportunity to fine tune the systems to maximise energy use, for example by resetting thermostats on hot water storage vessels to a slightly lower setting, provided that this does not affect the users unduly. Another example is when replacing windows and doors, to specify double or triple-glazed units, with appropriate draught sealing, for which the extra capital cost is marginal. However, a note of caution must be made regarding the overzealous introduction of draught sealing or similar measures in the name of energy conservation, as this may trigger or worsen other related environmental problems such as sick building syndrome or excessive condensation. In such cases, parallel measures such as efficient air filtration, trickle ventilation, or upgrading fabric insulation to avoid cold bridges may need to be taken as well.

11.10 Building environment

Over recent years increasing attention has been paid to the quality of the internal environment of buildings and its consequent effects on the health and comfort of the occupants. This has coincided with the move away from traditionally constructed and heated environments, particularly in the commercial sector, towards modern sealed and centrally serviced buildings, together with an increasing awareness of the deleterious effects on health which may be caused by such materials as asbestos, lead, and contaminated water. Writers on the subject such as Rostron[15] have commented on the 'high incidence of sickness amongst people who work in modern office buildings' which is 'not only of obvious concern to the sufferer but has commercial implications, in terms of increased absenteeism, reduced productivity, increased staff turnover, and low morale.' The term sick building syndrome (SBS) has been coined to describe the combination of causes that may trigger poor health in the users of such buildings. The symptoms, which vary widely according to the specific nature of the indoor environment and the susceptibility of the individual, can be broadly classed as follows:

- lethargy and loss of concentration
- headache, nausea and dizziness
- hoarseness, wheezing
- skin rash and itching
- eye and nose irritation
- eye strain
- backache and neckache.

Whereas some combinations of these symptoms may be found which have nothing to do with a building's internal environment, the presence of SBS can be further characterised by the following patterns:

- the symptoms disappear or decline away from work
- they are more prevalent in clerical staff
- they are more common in air-conditioned offices
- sufferers are likely to have little or no control over their work environment.

The causes are multiple, and the subtle interactions of several causes can obscure an accurate diagnosis in a particular instance. However, it would appear that the main cause is some forms of environmental stress, with the following factors being contributory to the syndrome:

(1)   Uncomfortable working environment as a result of:
- poor lighting
- poor posture at the workstation
- high or unsuitable temperature
- low air movement
(2)   Low relative humidity
(3)   Airborne odours, dusts and fibres
(4)   Chemical pollutants such as volatile organic compounds (VOCs).

Lack of personal control over ambient temperature, ventilation, lighting, and seating can exacerbate the symptoms, which is why SBS affects clerical staff disproportionately to those higher in the organisation who may be able to command an environment over which they have more control, for example by having their own office. However, there are a number of direct physical causes and triggers that the maintenance manager must address.

Ventilation and air quality

This is the single most important cause of SBS in air-conditioned buildings. Problems may stem from the low rate of air exchange in some air conditioning plant because of the perceived need to recycle air to minimise heat losses and thus increase energy efficiency. The average recommended ventilation rate of 1 air change per hour may need to be increased in buildings with a high level of airborne pollutants. Additionally, a low relative humidity may trigger symptoms such as dry throats and eye irritation.

The air itself may contain a range of dusts and pollutants that directly contribute towards SBS. In particular, dust from carpets and fabrics may linger in air-conditioned environments. The Building Research Establishment has carried out studies which indicate that steam cleaning of fabrics such as carpets and upholstery, plus a thorough vacuuming of hard surfaces to reduce dust levels, can lead to significant improvements in air quality and an amelioration of SBS symptoms in occupiers. Additionally, there are a number of harmful chemicals produced within the office environment (such as ozone from photocopiers and laser printers), and from outside (such as benzene from transport) which become concentrated to high levels because of the slow rate of air exchange in such environments [16].

Such pollutants can be reduced through the following measures:

- increase ventilation rates
- thorough and regular cleaning routines for internal building surfaces, and for ventilation systems
- siting of pollutant-generating equipment in closed, well-ventilated areas.

Temperature and humidity

Low humidity and unsuitable temperatures may lead to SBS symptoms. Humidity levels should preferably be in the range 40–60%, but it should be noted that humidifiers need regular maintenance to avoid contamination by microorganisms that may trigger legionnaires' disease (see below). ISO 7730 (1984) recommends that temperatures should be within the range 20–24°C, with a vertical differential between head and feet of less than 3°C, and mean air velocity of less than 0.15 m/s.

Lighting

Both the amount and variability of lighting levels are important to reduce eyestrain, particularly as computer VDU screens become more prevalent in the workplace. Low lighting levels and glare may cause headache and fatigue. Rapid variations in external light levels caused, for example, by sunshine and cloud, may cause strain where operatives need to concentrate on a particular task with a fixed level of lighting, such as a VDU screen. Glare and reflections caused through a poor combination of luminaires and surfaces may also trigger these symptoms. Flicker from malfunctioning fluorescent luminaires can be a particular irritant. The Health and Safety Executive publish a guidance booklet: *Lighting at Work* (1998 edition), which offers further advice in this respect, particularly relevant in view of the VDU screen regulations in force under the Health and Safety at Work Act.

Ergonomic factors

Sick building syndrome may be exacerbated by poor ergonomic design of the workplace. Excessive noise levels, for example in an open-plan office, may adversely affect the performance and comfort of workers. Poor workstation design which may cause a worker to stoop or lean forward to carry out tasks such as keying or reading a VDU screen can be major causes of backache, eyestrain, and repetitive strain injuries, with the subsequent discomfort and loss of productivity of the user and possible legal liability against the employer. The use of variable seating giving the operative a degree of control over posture is necessary. The Health and Safety Executive also publish a guidance booklet on seating at work (HSG 57).

To minimise the incidence of SBS, a check routine or audit of possible causes is desirable, such as is suggested by Rostron [15].

Legionnaires' disease

Legionnaires' disease is a deadly form of pneumonia caused by the *Legionella pneumophila* bacterium which may be found in contaminated water in ventilation humidifiers. It was identified in 1976 after a convention of veterans in Philadelphia led to the deaths of some of the participants. The bacteria thrive in temperatures of between 20°C and 45°C where there is water in spray or aerosol form, together with the presence of sludge or slime, timber, or other organic nutrient such as is commonly found in air conditioning systems which are poorly maintained. The disease can strike not only the users of the building, but also passers-by outside where they may become exposed to the air expelled from the air conditioning system.

The maintenance manager must ensure that adequate maintenance routines are in place to prevent infection of air conditioning and hot water systems by legionella, particularly in the case of complex or older systems where access to all possible breeding areas may be difficult because of the original design, or where the design is of such a nature that there may be areas at risk of contamination. Systems should be drained and disinfected at least twice a year to limit this risk. Biocides and disinfectants used may need to be varied to prevent colonisation by strains of the bacterium resistant to a particular agent. A detailed inspection of water towers, humidifiers, calorifiers, and storage tanks is necessary to check for damage or the build-up of sludge or slime. It is important to keep detailed records of these inspections to comply with Health and Safety at Work Act requirements relating to risk identification and monitoring (Chapter 3). Particular systems at risk are:

- systems with evaporative condensers
- systems with a cooling tower
- systems in buildings with susceptible occupants such as hospitals and surgeries
- humidifiers operating at or exceeding 20°C which creates water droplets
- baths and pools with agitated or circulated water
- systems with hot water services exceeding 300 L.

11.11 Asbestos and deleterious materials

There are a number of materials in use in existing buildings that have been found to have a deleterious effect on the environment and on people's health. Asbestos is the most commonly found, and potentially one of the most dangerous. It is a naturally occurring fibrous mineral with excellent fire resistant properties and binding properties, and was extensively used throughout the early part of the twentieth century as:

- insulation and lagging to heating systems
- fire-resistant boards
- an admixture to cement used in corrugated roof panels and flue liners
- in mastics and sealants as a binding agent
- in thermoplastic floor tiles.

Whilst all forms pose some threshold of hazard, it is its use in insulation and boards which poses the greatest risks. There are several forms of asbestos, the fibrous silicates: actinolite, amosite, anthophyllite, chrysotile, crocidolite and tremolite. Some, such as crocidolite, or blue asbestos, are particularly damaging to health even in low exposures. Asbestos fibres irritate the membranes of lung tissue, causing scarring and leading to emphysema and other irreversible respiratory diseases. It is therefore very important for the maintenance manager to be aware of the grave risks of exposure to asbestos fibres and to take action to avoid contamination. Exposure to asbestos fibres in the atmosphere is a particular risk, and therefore any lagging or board material which is discovered, damaged, or disturbed during either normal occupation of the building, or more usually, in carrying out maintenance works should be treated with extreme caution in case it does contain asbestos. The material must be left undisturbed, and all personnel removed from the area without delay; the area should be closed off pending further inspection by asbestos experts.

There are strict statutory controls governing those working on asbestos removal, contained in the Control of Asbestos at Work Regulations 1987 (SI 1987 No. 2115) and the Asbestos (Licensing) Regulations 1983 (SI 1983 No. 1649), which require operatives and companies carrying out asbestos removal work to be licensed by the Health and Safety Executive. In addition, the Health and Safety Executive must be notified on discovery of suspected asbestos. If found, the removal process is carefully controlled and will require the screening off of the affected area whilst removal is in progress. Air tests must be carried out frequently, and the affected area carefully vacuumed by the removal experts prior to the all-clear being given. The waste requires careful disposal at designated sites, and is strictly controlled by legislation.

Whereas asbestos can be very dangerous when discovered, because of the risk of airborne contamination by loose fibres, it is not considered dangerous when entrained with other materials in normal use. Therefore although great care should be taken by the maintenance manager and the maintenance operatives whilst carrying out investigations and work, there may be no need to remove asbestos if it is sealed and safe. If in doubt, expert opinion should be sought without delay. However, careful records should be kept of its location and the extent of its use, as outlined in the requirements for risk management in the workplace of the Health and Safety at Work Act.

Other materials in common use that may be now considered deleterious include lead used in plumbing, and some timber preservatives previously in use. On discovery, a risk assessment should be made by competent health and safety experts in accordance with the Health and Safety at Work Act and appropriate action taken when this has been done.

11.12 Procurement of materials, goods and services

Over recent years problems associated with environmental degradation as a result of the over-exploitation of natural resources have become manifest across the globe. Of particular concern to the maintenance manager is the use of materials such as tropical hardwood timber leading to the depletion of forest cover, and quarried and mined materials causing environmental damage and pollution. As part of an overall environmental policy, the procurement of goods, materials, and services used in buildings needs to be taken into account.

At the time of writing however, there is little guidance for the maintenance manager in the sourcing of environmentally responsible products, though this is slowly changing. For example, organisations such as the Timber Trades Federation and Friends of the Earth operate an eco-labelling system for timber drawn from sustainable sources. Other products such as chlorofluorocarbons (CFCs) previously used in refrigeration and air conditioning systems are being phased out because of damage to the ozone layer, though maintenance managers should note they may continue to be in use in some existing systems. Halon gas fire-fighting systems also have fallen out of favour because of their deleterious effect on the ozone layer when released.

Environmentally conscious specification of goods and services may nevertheless be implemented to some degree by maintenance managers. As providers of regular and repeat work to suppliers of goods and services, it is important that maintenance managers ensure their procurement practices encourage the environmentally responsible providers. In the absence of formal eco-labelling, the following agenda may be applied to the procurement process to assess the environmental impact of goods and services:

(1) minimal raw material consumption
(2) materials from sustainable/nondegrading sources
(3) minimal energy consumption
(4) minimal packaging
(5) incorporation of recycled materials
(6) potential for recycling
(7) locally sourced to minimise transportation pollution.

11.13 Pollution and waste management

Building users generates varying quantities of waste, both in the normal operation of the building (e.g. office waste, waste from production processes) and waste generated as a result of maintenance and alteration works to the building and its services. The maintenance manager will be concerned with both, in connection with its storage, and its disposal. The Environmental Protection Act 1990 lays a statutory duty on the producers, importers, holders or carriers of waste to:

● prevent the escape of waste
● secure proper disposal
● ensure transfer of controlled waste to a registered carrier.

In this respect 'controlled waste' includes most waste from building operations. A 'special' waste includes the more dangerous and polluting types of waste such as the asbestos referred to above, and further measures are necessary to ensure its safe disposal at approved sites.

In addition to the statutory requirement for the safe disposal and storage of wastes generated in a building, it may be decided as part of the occupier's environmental policy to recycle some or all of the waste generated. This will entail setting up housekeeping routines and separate bins to collect and store different types of waste, discriminating between glass, paper, organic, and other wastes, for example.

Other pollutants or nuisances generated by a building's use include emissions of smoke, noise, or dirt, all of which are controlled under the Environmental Protection Act 1990, described in more detail in Chapter 3.

References

(1) HRH the Prince of Wales (1989) *A Vision of Britain.* Doubleday, London.

(2) International Charter for the Conservation of Monuments and Sites (ICOMOS) (1981) the 'Venice Charter'.

(3) Ruskin, John (1849) *The Seven Lamps of Architecture.* Cassell, London.

(4) English Heritage (1991) *The Principles of Repair.* HMSO, London.

(5) Brand, Stewart (1994) *How Buildings Learn.* Viking Press, London.

(6) Morris, William, & Webb, Philip (1877) *SPAB Manifesto.* Society for the Protection of Ancient Buildings.

(7) Ruskin, John (1849) *The Seven Lamps of Architecture.* Cassell, London.

(8) English Heritage (1994) *Investigative Work on Historic Buildings.* HMSO, London.

(9) Harris, D.J., & Bowles, G. (1998) *Application of a Life Cycle Technique to the Environmental Assessment of Housing.* Royal Institution of Chartered Surveyors COBRA conference proceedings, Oxford, 1998, RICS, London.

(10) Department of the Environment (1991) *Environmental Action Guide for Building and Purchasing Managers.* HMSO, London. Ch. 3.

(11) British Standards (1992) *BS 7750: Specification for Environmental Management Systems.* British Standards Institute, London.

(12) Baldwin, R. for BREEAM (1998) *Environmental Assessment.* Building Research Establishment, Garston, Watford.

(13) Energy Efficiency Office (1991) *Energy Efficiency in Offices.* BRECSU, Garston.

(14) Royal Institution of Chartered Surveyors (1993) *Energy Appraisal of Existing Buildings.* RICS Books, London.

(15) Rostron, Jack (ed.) (1997) *Sick Building Syndrome: Concepts, Issues, and Practice.* E. & F.N. Spon, London.

(16) Perry, R *et al.* (1992) Indoor/Outdoor Air Quality Interactions in a Central London Office. Paper given at *CIB/W70 Conference: Management, Maintenance and Modernisation of Buildings.* Rotterdam, 1992.

Appendix
Standard Maintenance Descriptions

Standard Maintenance Descriptions (SMD) is a standard format for building maintenance information management, which can be adapted for use in any property management organisation. It is intended to parallel the Standard Method of Measurement as a common means by which information about building maintenance can be transmitted and shared. The format was published in 1992 by the Royal Institution of Chartered Surveyors as Research Papers 20.1 and 20.2, as a result of a proposal from the RICS Maintenance Committee (now the Building Maintenance Practice Panel). The research was carried out at Liverpool John Moores University and funded by the RICS Educational Trust.

Format for standard maintenance descriptions

Scope of SMD
Operational Requirements of SMD
SMD Format
SMD Data Categories

(1) Date and periodicity
(2) The asset register
(3) Building element description
(4) Property condition assessment
(5) Policy and planning
(6) Maintenance procurement
(7) Monitoring and record keeping

Framework for standard maintenance descriptions

Scope of SMD

The SMD format organises the information necessary for the planning and operation of the maintenance, alteration, and extension of buildings and their services and environment.

The essential components of these maintenance management operations are:

(1) Identification of maintenance need (predictive or reactive)
(2) Inspection to determine repair or servicing works required
(3) Policy and planning: priority and budgetary decision making
(4) Scheduling and implementing maintenance works
(5) Monitoring and payment of works carried out.

The SMD format is capable of expansion to include other operations, for example space planning, cleaning, security.

Operational requirements of SMD

(1) The format meets the operational needs of maintenance management planning and execution in a logical and consistent manner.
(2) The level of data detail is capable of expansion or contraction whilst retaining the format's structure.
(3) Data collection, processing, and storage needs to be kept to the minimum necessary to effectively operate the maintenance management process to deliver the desired standard of maintenance at the least cost.
(4) The format is directly applicable to a wide range of size and type of property estates.
(5) The format facilitates computerised data entry, processing, and retrieval.
(6) The format links with other data standards within the construction industry and generally, to provide a uniform data environment capable of import and export of data.

SMD format

The format consists of standard 'menus' for each category of information. The menus give a standard range of descriptions for the information in each category that will cover the great majority of circumstances encountered in building maintenance management. In each category there are also unused categories for user-defined fields.

The format is denoted by letter/number combinations. These are given as a means of shorthand reference. In normal use, it is envisaged that the full text description is used for clarity of understanding. These text descriptions are designed to be appropriate for *choice menus* in computer information-management software. The standardisation of choice menus renders data storage, analysis, and presentation capable of automation by macro program or dedicated software.

SMD data categories

(1) Date and periodicity

(2) The asset register
(3) Building element description
(4) Building element condition and functionality
(5) Criticality and priority indices
(6) Works scheduling
(7) Monitoring and record keeping.

(1) Date and periodicity

Time-related information in maintenance management, such as:

- dates of property acquisition or disposal
- dates of surveys
- priority and financial decisions relating to maintenance policy
- tendering and works ordering
- monitoring and updating

is recorded together with the date, in British calendar numerical format (day, month, year).

Example: **21/6/00** is 21 June 2000.

Cyclical operations, for example annual inspections and servicing, financial and accounting deadlines, or 5-year painting cycles, are recorded in parentheses after the date of the last or next such operation:

- **(y) year cycle.** *Example:* 21/6/00 (5y) is a 5-yearly inspection date.
- **(m) months cycle.** *Example:* 1/5/99 (1m) or (m) is a monthly cycle date.
- **(w) weekly cycle.** *Example:* 3/12/99 (w) is weekly; (2w) is fortnightly.

Time-related information (such as weather conditions during surveys) is recorded as a separate field linked to the date reference.

(2) The asset register

The coding system that follows can be used in whole or part to reference any building or area within the building for maintenance purposes.

   The full format is given, though this level of detail will not be necessary for many estates and uses. The format retains its structure however few or many of the categories are used.

   Each building and structure has an identity reference based on all or relevant parts of the following categories:

(a) Geographical location
(b) Predominant use(s) of the buildings or areas within buildings
(c) Building parameters: construction date(s), type, size and height

(d)   Tenancy.

(a) Geographical location
Consists of:

- Address reference to locate building or buildings complex
- Location reference to specify area within building(s).

Address
　　Pinpoint Address Code based on GIS Systems:
　　Name of Building Occupier
　　Postal address and Postcode
　　12-digit Ordnance Survey reference (Grid, Eastings, Northings)

Example:
　　Liverpool John Moores University
　　Clarence Street Building
　　Clarence Street
　　Liverpool L3 5UG
　　OS Reference: SJ 35412 90383

Location references within building/address
Uppercase letter followed by number:

- **B** = *Block* (building or group, if more than 1 building on PAC)
- **S** = *Section* (of block, if block is large or complex)
- **L** = *Level* (or storey) (e.g. 0 = ground, 1 = first, -2 = lower basement)
- **R** = *Room* (or area), numbered per block or section
- **E** = *Elevation* (by compass orientation e.g. North = 0, West = 270)
- **F** = *Roof* (if more than 1 roof area, append to Section, Level, or Room to specify location)

If there is only 1 building at the address, the Block reference does not need to be specified. In this case the default value of B is zero. The same applies to the Section reference S.

　　If the level is not specified, for example in a single storey building, then L = 0, equivalent to 'ground floor', is the default value.

　　Rooms or areas are numbered per block or section rather than per level so that staircases, balconies, atriums, or double storey rooms can be specified without ambiguity. Rooms subsequently partitioned or divided may be referenced by decimals as an alternative to adding nonsequential numbers or renumbering the whole building. For example Room 45, partitioned, may become 45.1, 45.2, 45.3, etc.

Example:
Block 2 Level 0 Room 34   is Room/area 34 on ground level of Block 2
Block1 Elevation 180   is the South elevation of Block 1
rooF Level 2   is the roof area over the 2nd storey of a single
   building (no Block reference required (or B = 0))

(b) Predominant use(s) of the buildings or areas within buildings

A use code reference to provide easy recognition of predominant uses, and to provide key for prioritising works based on the criticality/importance of the building/area/room use.

Use codes are the same at any level, to refer to the predominant use(s) of a whole building or block or to specific sections, rooms or areas within a building or block.

The use is specified as a lowercase letter following the block/section/level/room reference.

SMD reference		CI/SfB Table 0 equivalent
r	residential, single family	82
f	residential, flat/maisonette/shared use	81,84
n	residential, institutional	85
o	office or office area	32
s	shop or retail area	34
i	industrial/production area	26–28
q	official administrative areas (courts, town halls)	31
h	hospital or medical use	4–
a	assembly (including schoolrooms, churches, etc.)	6–,7–,91
d	exhibition/display areas (including museums, libraries)	75,76,78
p	pleasure or recreational use or area (e.g. cinema)	51–54,56
x	scientific equipment or computer area	3,78
e	restaurant or eating area, rest area	51,92
k	kitchen or food preparation area	3
w	sanitary accommodation	94
g	storage or warehousing	96
v	vehicle parking	12
m	plant or machinery rooms/areas	95,97
c	circulation areas	91
t	reception, waiting, or holding area	91
z	sleeping area	–

Five letters (b, j, l, u, y) are left free for allocation to individual estate uses, if the above classification is not sufficiently detailed. Combined or double letters may also be used, for example 'ke' for a kitchen/diner. The CI/SfB Table 0 equivalents are approximate only.

Example:

Block 1 office; Level 3 Room 45 wc	is a WC room on the 3rd floor of an office block
Block 1 office; Level 3 Room 44 office	is the office accommodation adjacent
residential Level 1 Room 6 z	is a bedroom of a 2 storey house
residential Room 3 k e	is a kitchen/diner on the ground floor of a house

(c) Building parameters: construction date(s), type, size and height
The date of construction and type of building may be recorded as separate entry fields linked to the Pinpoint Address Code or Block reference of the building.

- **The construction date** is merely the year of completion.
- **The construction type** is a brief outline phrase indicating walls or structural frame type, cladding, followed by/roof covering:
  Standard Construction Type phrases:

Structural frame,	**Cladding or walling/**	**Roof**
masonry	brick	p slate (p = pitched)
r.conc.	block	p tile
steel	conc. panel	p membrane (e.g. felt)
timber	render	f membrane (f = flat)
lightweight	timberclad	f asphalt
(other: specify)	stone (inc. slate)	p steel
	glass	
	(other: specify)	(other: specify)

Example:
Con.date: **1966**
Con.type: **r.conc., brick/f asphalt** (structure, cladding/roof)
Other information relating to:

- **size** (gross floor areas)
- **height** (number of storeys)

is recorded alongside a block and use code reference.

All information in these classes is enclosed in parentheses to differentiate between it and the location and use references above.

Example:

Gross floor area (in square metres), number of floors above ground, and number of basement floors of block 1 of a building:

B*lock 1(3500)* **L***evel (8,-2).*

or for a specific area,

Room 44 is an office of 345 m² floor area on the 3rd storey of the 8 storey office block with 2 basements in the example above:

B*lock 1 office(3500)* **L***evel 3(8,-2)* **R***oom 44 office (345)*

(d) Building tenancy

Where necessary to flag lease arrangements relating to maintenance planning and cost allocation, the Code may contain a **tenant reference** to indicate the type of tenancy and, if required, the name of the tenant.

The **tenant reference** is indicated by lowercase letter(s), and is in parentheses to distinguish from the location and use codes:

(f)   full repair obligation on tenant for whole building or facility
(p)   partial repair obligation on tenant for some items e.g. internal
(x)   no repair obligation on tenant
(s)   all or some repairs are charged as service charge to tenants
(c)   this area or element (e.g. elevation or roof) is common to several tenancies and chargeable to service charge.

Example:

Room 44 of the office block in the example above has the Tenant Reference:

B*lock 1 office* **L***evel 3* **R***oom 44 office* **(***ps, Smith & Co)*

to indicate room 44 is tenanted, partial repair plus service charge to common parts, lease held by Smith & Co.

(3), (4) Property condition assessment and surveys generally

Seven types of condition surveys are defined:

(1)   cursory inspections, broad-brush appraisals, and evaluations
(2)   structural condition surveys
(3)   cyclical surveys as part of a planned maintenance programme
(4)   specific detailed surveys to prioritise and schedule repairs
(5)   surveys to compare building, services provision and condition to user needs
(6)   specialist surveys such as dampness investigations
(7)   fault check or specific failure surveys

although a particular survey may combine several of these types.

The survey type should be chosen and structured to provide sufficient and sufficiently detailed information to fulfil the specific goals of the survey. Extraneous and irrelevant information should not be collected.

Condition surveys to an SMD format should be structured in such a way that data headings and terminology are consistent at any level. This will ensure economy in data collection time and cost, uniformity over time, and interchangeability of data within and between organisations.

(3) Building element descriptions

This classification is designed to provide the framework for defining and scheduling the forms of built construction and services within and around buildings. The element classification is linked to the location code to specify what the form of construction is, and where it is to be found. The structure enables any level of detail, from broad assessments of whole buildings or facades, to individual components of a building, to be recorded.

The system is structured around three main classes of data:

- **B: building fabric**
- **S: building and engineering services**
- **A: ancillary** – grounds, equipment, moveable items, ancillary services.

Elements and subelements within these classes are hierarchically defined to the requisite level of detail. A further lowercase letter gives location or other information e.g. external/internal, or whether non-structural.

The classification is given in outline, followed by the element groups expanded into greater detail in hierarchical form. Greater levels of detail, which may specify construction materials or manufacturers, are user-defined as required.

Building element descriptions outline element description classification
B: Building fabric:

- BN   foundations and structure
- BC   walling and cladding
- BW   windows, doors and lights
- BR   roofs and balconies
- BF   floors and ceilings
- BS   stairs and ramps
- BD   chimneys, shafts and ducts
- BCC   sundry fixtures

S: Service installations:

- SE   electrical systems
- SG   gas and other fuel services
- SW   hot and cold water services
- SV   mechanical ventilation and air conditioning systems
- SS   sanitary installations
- SD   drainage and refuse disposal
- ST   lifts and transportation
- SC   computer, communications and alarm systems.

A: Ancillary features:

- AA   roads, paths and access
- AF   fencing, plantings and external
- AS   sports facilities
- AX   other ancillary services.

Full element description classification

SMD format	CI/SfB Table 1 equivalent
B: BUILDING FABRIC	(1–) to (4–)
o: external	
i: internal	
u: underground or buried	
n: nonstructural or removable e.g. partition wall	
BN FOUNDATIONS and STRUCTURE	(1–) (2–)
Foundations	(1–)
strip, mass concrete	
raft	
pile and beam	
other	
Structure	(28)
framed	
panel	
mass	
tension	
floor/roof	
other	

BC   WALLING and CLADDING   (21) (41) (22) (42)
External
        structural masonry
        frame and panel
        cladding
        render and wall finishes
        coping and parapets
Internal
        solid
        frame/partition
        wall plaster and finishes
Sundry
        lintels and beams over openings
        built-in cills and labels
        damp-proof courses and seals
        external features and finishes
        skirting and rails
        finishes
        other

BW   WINDOWS, DOORS and LIGHTS   (31) (32)
Windows, etc.
        windows and frames
        glazed panels and lights
        rooflights and openings
        external windows and lights finishes
        internal windows and lights finishes
        other
Doors
        external
        internal
        large, garage, bay
        external finishes
        internal finishes
        other

BR   ROOFS and BALCONIES   (27) (37) (47)
Pitched   (27)
        structure
        covering
        insulation
        ventilation
        finishes

Flat
>   structure
>   covering
>   insulation
>   ventilation
>   finishes and solar reflective

Fixtures
>   gutters and channels
>   flashings
>   rails and guards
>   cradles
>   eaves and verges
>   eaves and verges finishes
>   other

Balconies
>   structure
>   surface/covering
>   finishes
>   other

BF   FLOORS and CEILINGS   (23)(33)(43)(35)(45)

Floors   (23)
>   structure
>   deck, boarding, screed
>   suspended floor
>   finishes and coverings
>   other

Ceilings
>   surface, deck, board
>   suspended
>   finishes and coverings
>   other

BS   STAIRS and RAMPS   (24) (34) (44)

Stairs
>   structure
>   rails and balustrades
>   finishes, coverings
>   other

Ramps
>   structure
>   rails and balustrades
>   finishes, coverings
>   other

BD   CHIMNEYS, SHAFTS and DUCTS   (2–) (3–) (4–)
Chimneys
    structure
    lining
    terminal
    opening/fireplace
    opening: finishes
    other
Shafts and Ducts
    structure
    openings
    linings and fixings
    other

BX   SUNDRY FIXTURES   (32) (33) (7–)
External
    canopies and awnings
    porches and covered walkways
    external features on building
    other external features
Internal
    blinds and curtains
    storage cupboards and shelves
    bars, counters, stages
    other internal features
Sundry
    other fixtures

S   SERVICE INSTALLATIONS
w: water or liquid-related
s: space or air-related
e: electric or electronic
d: cable, duct or channel
c: control, switch, or access point
t: terminal
a: appliance

SE   ELECTRICAL SYSTEMS   (6–)
    electrical intake and switchgear   (61)
    power   (62)
    lighting   (63)
    service to lifts, heating and a.c. supply, etc.   (5–) (6–)
    service to heating and a.c. etc.   (5–) (6–)
    service to communications   (64) (68)
    emergency and standby

external lighting and power
other

SG   GAS and OTHER FUEL SERVICES
Gas
    main, meters, stopvalve   (54)
    distribution system   (5–) (54)
    appliances
    other
Fuel Oil   (5–)
    storage
    distribution system
    appliances
    other
Solid Fuel
    storage
    distribution system
    appliances
    other
Solar Power and Other Energy Sources   (5–)
    storage
    distribution system
    appliances
    other

SW   HOT and COLD WATER SERVICES   (53)
Cold Water Supply
    water main, meters, stopcock
    cold water storage
    cold water distribution
    rising fire mains and sprinklers
    cold water: other
Hot Water
    hot water storage and calorifier
    hot water distribution
    hot water space heating distribution
    hot water space heating radiators

**SV   MECHANICAL VENTILATION and
AIR CONDITIONING SYSTEMS**   (57)
    mechanical ventilators
    air conditioning and treatment plant
    ventilation and air conditioning distribution
    refrigeration equipment
    mech. vent. and air conditioning: other

SS   SANITARY INSTALLATIONS   (52) (73) (74) (75)
Sanitary
     WCs
     baths and showers
     wash handbasins, sinks, and basins
     sundry fixtures
     traps and pipework
     other

SD   DRAINAGE and REFUSE DISPOSAL   (52)
Foul water
     above ground
     below ground
     sewage treatment
     sewer connections
Surface water
     above ground: gutters and downpipes
     below ground
     soakaways and land drains
Refuse disposal systems
     incinerators
     refuse chutes
     refuse storage
     refuse: other

ST   LIFTS and TRANSPORTATION   (66)
Passenger lifts
     mechanism
     car
     finishes
Goods hoists and winches
     mechanism
     car
Escalators
     mechanism
     walkway
     finishes
Window Cleaning Cradles, etc.
Transportation: other

SC   COMPUTER, COMMUNICATIONS and ALARM SYSTEMS
     telecommunications intakes   (64)
     telecommunications
     fire defence systems
     security systems
     computer networks and data links
     audio visual systems
     CC and A: other

A: **ANCILLARY**   (90)
AA   ROADS, PATHS and ACCESS
     vehicle access: roads and car parks
     pedestrian access: paths, paved areas
     gates and barriers
     external signs and notices
     access: other

AF   FENCING, PLANTINGS and EXTERNAL   a
     fences, walls and fixed barriers
     external features
     external facilities (e.g. sheds and stores)
     lawns borders and plantings
     water features
     FP and E: other

AS   SPORTS FACILITIES
     pavilions, changing rooms and stores
     stands and external seating
     open games courts
     open sports areas
     pools and water sports features
     sports facilities: other

AX   OTHER ANCILLARY SERVICES
     moveable plant and equipment
     other
     This section can be expanded as required to include the wider needs of
     Facilities Management or particular users e.g. manufacturing, storage,
     retail, etc., or interfaced with other information systems relevant to the
     maintenance of any fixed or moveable asset.

These elemental divisions enable the building to be described as a whole, or to
be referenced to the Block/Level/Room codes described in the Asset Register
(Section 4.1).

(4) Property condition assessment

Condition criteria are defined in three categories:

- **physical condition**
- **obsolescence**
- **environmental assessment**

according to the standard alphanumeric grading index detailed below.

Not all three categories will necessarily be relevant in any one inspection process.

Condition criteria may describe a particular element, or a group of elements, such as those comprising an elevation of a building.

The criteria accorded to the element's condition may be accompanied by a specific description of the defect, and the repair or other action required, if the criteria grading is such that the repair is likely to be done before the next inspection.

Physical condition

This criterion assesses the physical state of repair of the element.

- **C0:** not inspected/recorded; assumed satisfactory
- **C1:** as-new condition
- **C2:** fully functional, showing some signs of wear
- **C3:** function slightly impaired in some ways, but still operational
- **C4:** function deteriorating and requiring attention soon
- **C5:** nonfunctional or absent
- **C6:** dangerous.

The C1 'as new' category is the starting point for a new element in a new building, from which point subsequent deterioration is measured.

The C3 and C4 assessments may be suffixed by / and a number indicating the time in years to anticipated failure, e.g. **C4/1** = within 1 year

'R' may be suffixed to indicate replacement as only option (i.e. the element cannot be repaired in situ) e.g. **C4R/1**.

The C6 category is intended as a specific highlight for very urgent or emergency work, on which immediate action is required.

The C0–C6 levels are sufficiently specific for recording most maintenance needs. However individual users may specify the codes in more detail, using a decimal point, if a finer 'grain' is required.

For example, **C5.1** may be nonfunctional; **C5.2** absent.

Obsolescence

This criterion assesses user need versus current provision.

- **N0:** not inspected/recorded; assumed satisfactory
- **N1:** element satisfactory for present use
- **N2:** element functional, but falls short of new product/installation
- **N3:** element is not adequate for user needs by current standards
- **N4:** element is wholly inadequate for users and seriously affects their use of the premises.

The N0 has a zero value as a default, i.e. if no entry is made the element is assumed to be satisfactory.

The N3 and N4 classes may require some additional detail as to the nature of the shortcoming if it is intended to raise works orders direct from the survey.

Environmental assessment
This criterion assesses the environmental impact of an element in use.

- **E0:** not inspected/recorded; assumed satisfactory
- **E1:** element's environmental impact satisfactory for present use
- **E2:** element's environmental impact acceptable, but falls short of new installation
- **E3:** element's environmental impact is not adequate for user needs by current standards, and detracts from the use of the building
- **E4:** element's environmental impact contravenes current legislative standards.

Thus, the action threshold will usually be at least E4, but preferably E3.

Some users may prefer to subsume the Environmental Assessment Criteria within the Obsolescence Criteria to minimise data categories.

(5) Policy and planning
The Index of Criticality is a classification of the levels of maintenance appropriate to particular areas or buildings according to their relative importance to the building user.

There are 2 components:

- the **location** component
- the **use** component

Both are rated from 1 to 5.

Location index
Level **L1** – (very high)	operating theatres, computer areas
Level **L2** – (high)	boardrooms, reception, prestige areas
Level **L3** – (standard)	main office areas
Level **L4** – (low)	sheds, outbuildings, low-grade ancillary
Level **L5** – (very low)	prior to demolition; short-life

Use index
Level **U1** – (very high)   health and safety related, fire defence, etc.
Level **U2** – (high)   failure would halt user operations
Level **U3** – (standard)   failure would be costly, or inhibit users
Level **U4** – (low)   failure would be moderately costly, etc.
Level **U5** – (very low)   failure would be a minor nuisance

The **Combined Criticality Index** (CCI), based on the importance and status of the element or area within the estate, plus the consequences of element failure and/or cost growth of repair postponement, is arrived at by multiplying the Location and Use Indices, giving a range of 1–25. For example:

The **CCI** of a broken window in a main office area could be: **L3** (standard location) × **U1** (health & safety hazard) = **3**.

The **CCI** of a leaking gutter in a garage could be: **L4** (low-grade location) × **U4** (moderate nuisance) = **16**.

The weightings of the location and use indices, here given as 1:1, may be varied according to the needs and priorities of the particular user.

Action thresholds and priorities may then be assigned according to the CCI. The lower action threshold CCI may be set at 15, i.e., anything greater gets actioned as and when resources are available.

This CCI, coupled with the observed state of repair of the element, provides the basis for prioritising repair requirements for a property estate, and could be determined according to the particular requirements of the maintenance organisation at that time without compromising the standardisation or quality of the condition data collected.

(6) Maintenance procurement

The repair schedule will consist of a description of the element and its location, plus a 'command' phrase consisting of the operation(s) to be performed and the materials needed.

Maintenance operations
General:

(1) Travel to site, attendance, screening and user protection
(2) Supply only goods or equipment
(3) Test, check, and/or report
(4) Routine servicing (e.g. to lifts, boilers, may even include grass cutting and gully cleaning).

Maintenance works:

(5) Patch repair or piece in
(6) Replace whole element or installation

(7) Preservation, decoration, or coating operation (e.g. painting)
(8) Cleaning or removal operation

Alteration and improvement works:

(9) Add new installation/element (e.g. new electrical circuit)
(10) Upgrade existing element to current standards (e.g. telephone system)

Example:
An instruction to a contractor to investigate and repair a leaking window frame to an office block may be coded and computer generated from maintenance records as follows:

Location:	Clarence Street Building, Clarence Street, Liverpool L3 5UG
	OS Ref: SJ 35412 90383
	***B**lock 1 (8 storey), West **E**levation (270)*
	*Level 2 **R**oom 201 office: At 2nd floor level to Room 201 (an office).*
Element:	*Building fabric **(B)***
	*outside and inside affected **(oi)**,*
	*Window **(FW)***
	opening frame *affected (window or door)*
Repair Operations:	*[1] Attend site*
	[3] Test/check/report
	[5] Patch repair
Command Phrase:	*'**Rectify water ingress**'*
Date:	*21/10/99*
CCI Index:	*L3 × U4 = **12** (to be undertaken within 3 months)*

All data and phrasing is automatically generated from the survey data or repair request log by single keystroke codes except the command phrase. The CCI index cues a response time request to the contractor. The CCI combined with the date reference provides a bring-up mechanism to monitor whether the repair has been attended to.

(7) Monitoring and record keeping
Updating Records
On completion of maintenance or servicing works, *whether inspected or not,* the property records must be systematically updated. Indication should be made of:

(1) date of completion of works
(2) whether the completed works were inspected

(3) details of new components or elements installed

(4) whether any works are still outstanding.

This would be best done on clearance of the invoice for payment.

This mechanism can also be used to monitor contractor/operative performance. Changes in the property portfolio relating to acquisition and disposal of buildings, and changes of use or occupancy, must also systematically be entered into the maintenance management system.

Further Reading and Contacts

General reference

Planned Building Maintenance – A Guidance Note
Royal Institution of Chartered Surveyors (1990) RICS, London.

Building Maintenance Management
Chanter, B. & Swallow, P. (1996) Blackwell Science, Oxford.

Managing Building Maintenance
Harlow, P. (ed) (1984) Chartered Institute of Building, Ascot.

Maintenance Management – A Guide to Good Practice
CIOB (1990) Chartered Institute of Building, Ascot.

Building Maintenance
Seeley, I. (1987) (2nd ed) Macmillan, Basingstoke.

How Buildings Learn
Brand, S. (1994) Viking, London.

Sick Building Syndrome – Concepts Issues and Practice
Rostron, J. (ed) (1997) E. & F.N. Spon, London.

The Economic Significance of Maintenance
BMI Special Report 264 (1997) Building Maintenance Information, Kingston.

Review of Maintenance Procurement Practice
BMI Special Report 270 (1998) Building Maintenance Information, Kingston.

Procurement of Maintenance Works – A Strategy
Royal Institution of Chartered Surveyors (1999) RICS Books, London.

BS 3811: 1984 British Standard Glossary of Maintenance Terms in Terotechnology
British Standards Institute, London.

BS 8210: 1986 British Standard Guide to Building Maintenance Management
British Standards Institute, London.

Environmental Action Guide for Building & Purchasing Managers
Department of Environment (1991) HMSO, London.

Building Maintenance & Preservation
Mills, E.D. (1994) Butterworth-Heinemann, London.

Facilities management

Facilities Economics
Williams, B. (1994) Bernard Williams Associates, BEB, Bromley.

Facilities Management – Towards Best Practice
Barrett, P. (ed) (1995) Blackwell Science, Oxford.

Facilities Management – An Explanation
Park, A. (1994) Macmillan, Basingstoke.

Facilities Management – Theory and Practice
Alexander, K. (1996) E. & F.N. Spon, London.

Disaster Planning and Recovery – A Guide for Facilities Professionals
Levitt, A. (1997) John Wiley & Son, New York.

Costing

Building Maintenance Price Book
Building Maintenance Information Ltd (Quarterly) BMI, London.

Review of Maintenance Costs
BMI Special Report 272 (1998) Building Maintenance Information, London.

Accounting and Finance for Building and Surveying
Jennings, A.R. (1995) Macmillan, Basingstoke.

Collection and Use of Building Maintenance Cost Data
Building Research Establishment (1978) BRE, Watford.

Building Costs, Use and Management
Grimshaw, B. & Nutt, B. (eds) (1999) E. & F.N. Spon, London.

Condition surveys

Condition Surveys: Building Maintenance
Bathurst, P. (1988) BMI Serial 167, BMI, London.

Stock Condition Surveys – A Basic Guide for Housing Associations
Hellyer, B. & Mayer, P. (1994) National Federation of Housing Associations, London.

Condition Assessment Surveys
Bobbett, I. (1995) *Building Maintenance Information Special Report*, BMI, London.

Energy Appraisal of Existing Buildings
Royal Institution of Chartered Surveyors (1993) RICS Books, London.

Building Surveys and Inspections of Commercial & Industrial Property – a Guidance Note for Surveyors
Royal Institution of Chartered Surveyors (1998) RICS Books, London.

Building Surveys of Residential Property – a Guidance Note for Surveyors
Royal Institution of Chartered Surveyors (1986) RICS Books, London.

Stock Condition Surveys – A Guidance Note
Royal Institution of Chartered Surveyors (1997) RICS Books, London.

Surveying Historic Buildings
Watt, D. & Swallow, P. (1996) Donhead Publishing, London.

Maintenance technology

Building Maintenance Technology
Son, L.H. & Yuen, G. (1993) Macmillan Press, Oxford.

Mitchell's Materials
Everett., A. (1994) Longman, Harlow.

Dampness in Buildings
Oliver, A., Douglas, J. & Stirling, J. Stewart (1997) Blackwell Science, Oxford.

Technology of Building Defects
Hinks, J. & Cook, G. (1997) E. & F.N. Spon, London.

Understanding Housing Defects
Marchall, Worthing, & Heath (1998) Estates Gazette, London.

Repair and Maintenance of Houses
Melville, I. & Gordon, G. (1997) (2nd edn) Estates Gazette, London.

Component lifespans

HAPM Component Life Manual
Housing Association Property Mutual (1995) E. & F.N. Spon, London.

Life Expectancies of Building Components
Building Surveying Division Research Group (1992) RICS, London.

BS 7543: 1992: Guide to the Durability of Buildings and Buildings Elements, Products, and Components
British Standards Institute, London.

Maintenance Cycles and Life Expectancies of Building Materials and Components – A Guide to Data and Sources
NBA Construction Consultants (1985) NBACC, London.

Information technology and data handling

Standard Maintenance Descriptions:
Part 1: Maintenance Management Practice (Paper 20.1)
Part 2: the SMD Format (Paper 20.2)
Wordsworth, P. (1992) Royal Institution of Chartered Surveyors, London.

Computer Aided Maintenance and Facilities Management Systems for Building Maintenance
BMI Special Report No. 260 June (1997) BMI, London.

Information and Data Modelling
Benyon, D. (1990) Blackwell Science, Oxford.

Value Management in Design & Construction
Kelly, J. & Male, S. (1993) E. & F.N. Spon, London.

From Data to Database
Bowers, D.S. (1988) Van Nostrand Reinhold, London.

Useful addresses and contacts

Building Maintenance Information Ltd
12 Great George Street Parliament Square London SW1P 3AD
Tel: 020 7222 7000, 020 8546 7555

British Board of Agrément
PO Box 195 Bucknalls Lane Garston Watford Hertfordshire WD2 7NG
Tel: 01923 670844; Fax: 01923 662133

British Institute of Facilities Management
67 High Street Saffron Walden Essex CB10 1AA
Tel: 01799 508608; Fax: 01799 513237

British Standards Institution
389 Chiswick High Road London W4 4AL
Tel: 020 8629 9000; Fax: 020 8996 7400

Building Research Establishment
Bucknall Lane Garston Watford Hertfordshire WD2 7JR
Tel: 01923 664000; Fax: 01923 664010
bre.co.uk

Building Services Research and Information Association
Old Bracknell Lane West Bracknell Berkshire RG12 7AH
Tel: 01344 426511; Fax: 01344 487575

Chartered Institute of Building
Englemere Kings Ride Ascot Berkshire SL5 7TB
Tel: 01344 630700; Fax: 01344 630777

Department of the Environment, Transport and the Regions (Planning Policy
Directorate, Construction Directorate and Building Regulations Division)
2 Marsham Street London SW1P 3EB
Tel: 020 7276 6645
http://www.detr.gov.uk

English Heritage
23 Savile Row London W1X 1AB
Tel: 020 7973 3000; Fax: 020 7973 3001

Health and Safety Executive
Rose Court 2 Southwark Bridge London SE1 9HS
Tel: 020 7717 6000; Fax: 020 7717 6717

Institute of Maintenance and Building Management
Keets House 30 East Street Farnham Surrey GU9 7SW
Tel: 01252 710994; Fax: 01252 737741

Joint Contracts Tribunal
88 New Cavendish Street London W1M 8AD
Tel: 020 7580 5588

Royal Institution of Chartered Surveyors
12 Great George Street London SW1P 3AD
Tel: 020 7222 7000; Fax: 020 7222 9430

Index